CITY ASTRONOMY

ROBIN SCAGELL

Series editor Leif J. Robinson

Sky Publishing Corporation
Cambridge, Massachusetts

Published in the United States by Sky Publishing Corp.,
49 Bay State Road, Cambridge, MA 02138

First published in Great Britain in 1994
by George Philip Limited,
an imprint of Reed Consumer Books Limited,
Michelin House, 81 Fulham Road, London SW3 6RB
and Auckland, Melbourne, Singapore, and Toronto

ISBN 0-933346-75-1

Edited by John Woodruff
Illustrations by Raymond Turvey
Page design by Jessica Caws
Produced by Mandarin Offset
Printed in Hong Kong

Contents

Foreword

Not much attention has been paid to sky-gazers who live in urban areas, where the stars are diminished by pollution from smoke-stacks, automobiles, and artificial lights. Many people, I fear, believe you can have fun with astronomy only if your skies are dark and pristine, like those above mountaintops. After all, aren't those the places where the world's major observatories are located? In this second volume of our series on observational amateur astronomy, Robin Scagell takes us on a voyage through blemished skies, demonstrating how to make the best of poor conditions.

From your backyard or rooftop observing site, your night sky will be illuminated by light from the surrounding city or town. And, like everyone else, you will have to contend with the weather, but you will have the extra problem of poor air quality. You may, for example, have to wait for a cold front to clean the skies from humidity haze. But there is still a host of celestial delights to be seen. Although your unaided eye may be able to pick out only a few hundred stars, or even just a few score, binoculars or a small telescope will reveal many times those numbers. A little optical aid can also give you good views of every type of major astronomical object – including star clusters, nebulae, and galaxies.

For those who want more, technology has recently come to the rescue. For visual observers there are special filters that let through the light from distant nebulae while blocking out wavelengths infested by unwanted stray light from streetlights. And modern solid-state chips called charged-coupled devices, or CCDs, allow modest amateur telescopes to penetrate the urban sky glow and reveal sights that would have taxed the largest professional instruments only twenty years ago.

I remember finding my first telescopic object: the double star Albireo in the head of Cygnus, the Swan. As a child, with my hand-held spyglass propped on the staircase, I could see the components' contrasting blue and yellow hues. The thrill of such discovery would be no less today in even the most light-polluted sky. The Double Cluster in Perseus still beckons this suburbanite, as does the Great Galaxy in Andromeda. City skies do not prevent us from tracking the red giant star Mira as it brightens from obscurity, only to fade back again months later. Have you ever looked at the Moon through binoculars or a small telescope – I mean *really* looked at it? And have you watched the mathematical dance of Jupiter's four bright satellites?

These celestial sights, and scores more, await the patient and intrepid connoisseur of the heavens. Robin Scagell's guide will let you get the most out of almost any sky with whatever equipment you have, or even with none at all.

Leif J. Robinson
Editor, *Sky & Telescope* magazine

Preface

These days, you don't have to live close to a city centre to suffer from city skies. At night, a pall of light spreads far beyond the sylvan suburbs, robbing us of a sight that everyone who lived before modern times took for granted – the spectacle of the heavens. The problem is called light pollution: just as a contaminant may pollute a river, so nighttime lighting is polluting our skies. It has increased dramatically in recent years, to the extent that even non-astronomers comment on it. Its effect on the skies may be obvious, but it is not destructive in the same way that river pollution poisons life. And, as I aim to show, it can be combated.

In the course of writing this book, I have been in touch with amateur astronomers in many parts of the world. It became clear that we are all in the same situation, no matter where we live. It is true that, while I observe from my *back garden* and use a red *torch* when recording my observations, an American will be in the *backyard* and using a red *flashlight* for the same purpose, but there the differences end. We are all faced with skies so bright that we hardly need the torch/flashlight anyway.

This book is as much for relative newcomers to astronomy as it is for those who have been observing for years. I have assumed that the reader is familiar with some of the basics of amateur astronomy, so I do not go into the details of standard observing techniques or terminology. If you don't know what an equatorial mount is, or whether first-magnitude stars are brighter or fainter than sixth-magnitude ones, then please look in a manual of amateur astronomy – some are mentioned in the Bibliography. At the same time I have tried to make the text as understandable as possible to the newcomer, but to explain the full implications of every term used would have turned this into a general handbook on amateur astronomy, which was not my intention.

Thanks are due to the many people who helped me compile this book by sharing their experiences and in some cases taking photographs specially. They are generally credited in the text. Leif Robinson's comments on the first draft were particularly welcome. In addition my special thanks go to John Woodruff for his editing skills, which combine a knowledge of astronomy with a sharp eye for errant text.

I hope this book will stimulate you to go out and actually look at the objects described – or maybe help you regain the enthusiasm that deserted you when the night sky got brighter. The good news is that the Universe is still out there and waiting to enthral you.

Robin Scagell
Vice-President, Society for Popular Astronomy

CHAPTER 1

Introduction

A clear blue evening sky is a delight, whether you live in the heart of a big city or in the depths of the country. The Sun has set, the first stars are starting to appear high up, and there is the promise of a fine night's observing. This is the time when every astronomer feels the tug of the stars.

By "astronomer," I mean anyone who wants to be one: you do not need to be a professional, or even a committed amateur with an observatory full of the latest equipment. It is possible to be an astronomer at heart without ever thinking of owning a telescope. Anyone who has gazed at the stars and has wished they could put names to what they see has felt the stirrings of astronomy. After all, the word "astronomer" simply means "star namer." It would be quite easy to make up your own names for the patterns of stars – there is an obvious saucepan; that group of stars below it looks just like a cat sitting on a fence; there is a kite, over to the left. But others have already given these patterns names, so it is those that we set out to learn instead: the ancient astronomers saw a bear, a lion, and a herdsman in the stars.

It is a strange feeling, as you look up, to realize that those stars glimmering in our modern electric sky with its unnatural pink glow are the same ones the ancient stargazers looked upon and named. Yet there is a world of difference between what most of us can see and what they could see. If you live in a city or a town, or out in the suburbs, the promise offered by that clear blue twilight sky is usually a false one. The wonder of stargazing has been taken from us by the need of millions of people to drive, to work, or to be entertained at nighttime, and to feel safe from nocturnal criminals. The artificial lighting that makes all this possible sends light up into the night sky, producing what is called *light pollution*. Figure 1.1 shows how light pollution affects our view of the stars, and Figure 1.2 shows how it makes the Earth look from space.

I think it would be too much to hope for that, as civilization progresses, we can regain the truly dark night sky our ancestors enjoyed. (Although fortunately – given the will to do it – the blight of light pollution is instantly reversible, unlike other forms of pollution.) Instead, as astronomers, our best course is to make the most of what we can see. For one thing, it can actually be easier to familiarize yourself with the sky from the city than from a dark country site. Most of the constellation patterns are framed around the brighter stars, which in a light-polluted sky are all you can see. Even those who know their way around the constellations can get confused when they find themselves under a perfect sky, with a myriad of stars jostling for attention.

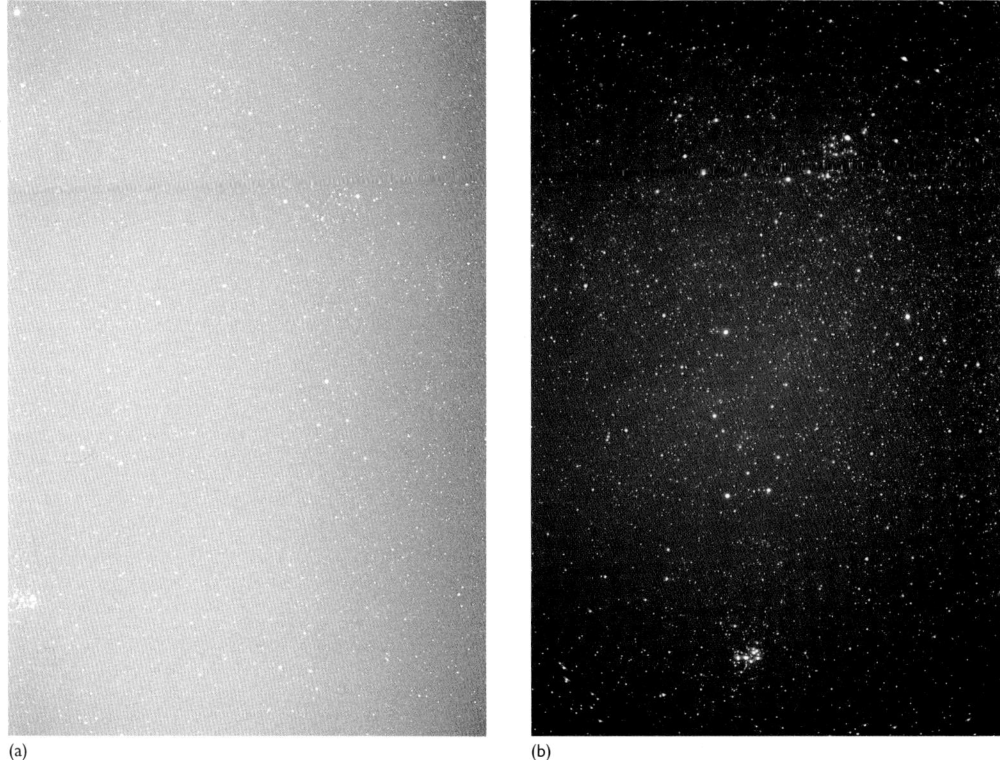
(a) (b)

Figure 1.1 *Two photographs of the Perseus region show the effects of light pollution. From the suburbs (a), the sky brightness drowns the faint stars. From a country site (b), however, a longer exposure is possible, recording fainter stars and the California Nebula (the reddish patch to centre left).*

Most suburban skies show only stars brighter than magnitude 4 or 4.5, even at the zenith. There are fewer than 1000 stars in the entire sky brighter than magnitude 4.5, of which about 500 are above the horizon at any one time (actually slightly fewer in the northern hemisphere, slightly more in the southern hemisphere). Of these 500, many will be so low in the sky that they are either not visible, or are hidden by buildings, trees, or other obstructions along the horizon. So, out of the 9000 or so stars that are visible to the naked eye in the entire sky under ideal conditions, you can probably see only about 200 at any one time from your home.

While this makes the task of learning the sky very much easier, without the allure of a star-filled night the invitation to explore further is nothing like as powerful. This, I believe, is the main reason why fewer people actually observe today than in former times. With modern techniques, as we shall see, it is still possible to observe a large proportion of the objects you might want to see. And with the advent of new electronic imaging devices, amateur astronomers can reach farther into the universe than ever before, even with the handicap of light-polluted skies. But there is little doubt that a hazy pink sky does not promise as much as a dark, crystal clear one. It is as inviting to astronomers as a murky pool is to swimmers.

The poor skies also play their part in dampening any initial enthusiasm for astronomy. However, even with a cheap pair of binoculars you can

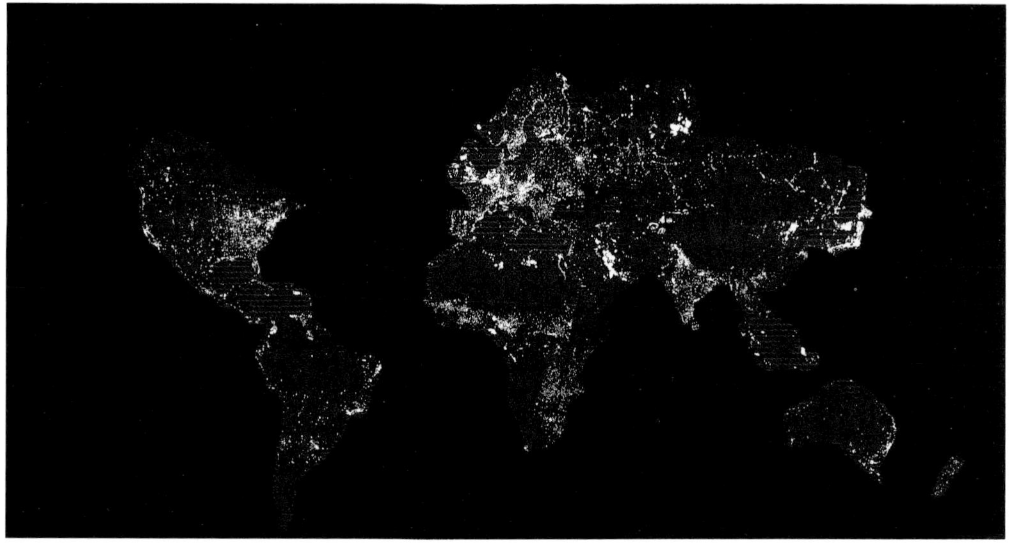

Figure 1.2 *A composite image of the Earth by night from satellite pictures. The glare from millions of lights spills upwards into space, creating visible sky glow which outlines each continent. Although the intensity of light generally indicates population density, there are exceptions – France, despite its affluent population, is darker than its European neighbours, for example. Gas flares from oil fields are prominent in the Middle East and elsewhere, while brilliant lights in the Sea of Japan mark the position of squid-fishing fleets which use light to attract their prey.*
© 1985 Woodruff T. Sullivan III.

locate many celestial objects that are invisible to the naked eye – although, sadly, few people ever make the attempt. The message of this book is that from an urban site you can still see a lot without having to spend a lot. As for the *types* of object you might wish to observe, such as planets, stars, or nebulae, about two-thirds of them can be observed from the city, including most of the spectacular examples. All you really need when confronted with bright night skies is enthusiasm.

Serious astronomy?

Some astronomers are a little like those hardy backpackers you come across in country areas who make you feel guilty about using your car to drive down the lane rather than walk. Just as the backpackers seem to believe that unless you are suffering you might as well not bother, those astronomers would say that astronomy is worth doing only if there is a result at the end of it. Serious observing is the important thing: if you are not making a contribution to knowledge, you are wasting your time.

But you can enjoy the countryside through your car window if that is what you want. Strolling down the lane is fun, too, and just because you never go more than half a mile from the car you will not be enjoying it any the less. Similarly, just gazing at the showpieces of the sky is rewarding, even if they are the same ones you saw last week or last month or last year. You should not feel guilty for not pushing forward the frontiers of science.

On the other hand, if you would like to have some sort of goal in mind then there are lines of work you can pursue from the city just as effectively as from the country. Amateurs like to believe that what they do is scientifically useful, and indeed it can be. There was a time, often called the Golden Age of amateur observing, when amateurs could make genuine discoveries. The eminent Victorian observer William Lassell, for example, discovered several planetary satellites, Triton among them, while making his money as a Liverpool brewer. During this century, until the Space Age, amateurs played a leading role in monitoring the Moon and planets. This first Golden Age is definitely over, though work in some traditional amateur fields, such as variable star observing, has not become outmoded despite the advent of improved observing methods.

But for the serious amateur a new Golden Age is just beginning. At one time, many amateurs had to make their own specialized equipment. These days, there is a wider range available off the shelf and at affordable prices than ever before. And with the arrival of electronic imaging, the well-equipped amateur is now in a position to carry out types of work, such as astrometry, that were previously virtually out of reach. With a comparatively modest outlay, an enterprising amateur can now enter fields that were once the exclusive preserve of professionals.

The cost

What can you see, and what will it cost? Apart from the fun of learning the sky, which costs virtually nothing, you can get a long way with nothing more than a pair of binoculars costing about the same as a budget camera. These will show you, for example, the more prominent star clusters and nebulae, even from badly light-polluted skies, as well as bright comets. If you want to make useful observations, you can tackle variable stars and even meteors.

The next stage up is to acquire a telescope. Be extremely wary of very small telescopes of the sort offered in mail-order catalogues or by non-astronomical suppliers such as department stores and camera shops. Although some of them are quite suitable for the beginner, most are of poor quality, and by the time you find you have bought a lemon there is little you can do about it. The telescope market now offers a wonderful variety of instruments to suit all pockets, sometimes literally, but I still believe that a good 150-mm (6-inch) reflector, like the one shown in Figure 1.3, is very hard to beat for value and usefulness. You can buy a basic instrument for about the price of a reasonable video recorder.

With such a telescope you can see considerable detail on the Moon, pick out markings on the bright planets, and get good views of some 400–600 deep-sky objects, even from badly light-polluted areas. With the right mounting, drive, and filters, you can also take impressive photographs that will show few signs of the effects of light pollution.

You can, if you wish, spend huge sums on telescopes of increasing sophistication. A reasonable next step up might be a fairly large Dobsonian reflector, with which you can observe fainter deep-sky objects; or maybe a well-mounted reflector of 250–300 mm (10–12 inches) aperture, so that you can photograph a wider range of subjects. This would cost about the same as you might spend on a good used car. A very important addition to the amateur astronomer's armoury is the charge-coupled device, better known as the CCD. This enables you to obtain amazing images of objects that cannot be recorded by conventional photography. A typical CCD camera costs about as much as the computer you need to display and

Figure 1.3 *A basic 150-mm (6-inch) Newtonian reflector, here shown on an altazimuth mount, is an excellent compromise between portability and economy. This instrument will outperform smaller refractors with fluorite optics costing several times as much.*

store the CCD images (not to mention the permanent observatory of some sort that you will need to keep the equipment safe).

No matter how much you spend, you will never be able to defeat the streetlights completely. The naked-eye spectacle of the Milky Way can now be enjoyed only well away from urban areas, and in some cases only by travelling abroad. Similarly, from the city deep-sky objects such as clusters and galaxies will rarely match the descriptions you will find in observing guides. Most galaxies are either invisible or merely ghosts of what they are like in truly dark skies. Your own wide-field colour photographs of the heavens are also likely to be a disappointment. Chapter 7 deals with how to find those dark skies that will allow you to observe the objects that are otherwise hidden from you.

CHAPTER 2

Know your enemy – the weather

Astronomers are at the mercy of the weather. The new observer soon finds that it has plenty of tricks up its sleeve. Urban astronomers in particular need to keep a keen eye open, as they have not only the weather to contend with but also the light pollution. The two are linked, as we shall see.

When it comes to the effect of the weather on observing conditions, myths abound. Several of these I aim to correct:

1 Light pollution is the only reason why the night sky is no longer dark.
2 Rain washes the dirt out of the air.
3 Mist creates good seeing.
4 In an unreliable climate there's no point in doing astronomy.

In the course of this chapter I shall demolish each of these myths in turn, and show how you can improve your observations, even from the city.

The role of water vapour

Light pollution is *not* the only reason why the night sky is no longer dark. Have you ever wondered what the city lights are shining on? The atmosphere contains more than just air. It carries gases, aerosols, and dust particles from industry, aircraft, wind-blown soil, forest fires, volcanoes, and meteoroids, maybe pollen grains, and, above all, water vapour. Except over certain areas, the effects on sky clarity of all the others are usually trivial compared with that of water vapour.

Water vapour is different from water droplets or steam. It is the gaseous form of water, and it behaves like a gas. Even when the temperature is well below freezing there can still be water vapour in the atmosphere, though warm air can hold very much more water vapour than cold air can. That is why washing on the line dries best on a warm day. Boiling is a quick way of converting water to vapour, but water will still evaporate at lower temperatures – a wet road does not need to be heated to the boiling point of water in order for it to dry out. Even on a freezing day, wet objects will eventually dry, showing that water can still evaporate in subzero temperatures.

The amount of water vapour in the atmosphere is measured by the humidity. (More specifically, it is measured by the *relative humidity*, the ratio of the actual water vapour content to the maximum possible content.) From the astronomer's point of view, the humidity is an important quantity: high humidity means milky skies during the day and

(a)

(b)

Figure 2.1 *(a) A clear day in Britain, with the sky milky as a result of water vapour in the atmosphere. This is also common throughout much of Europe and the US. (b) The sky over the Mills Cross radio telescope at Molonglo near Canberra in Australia, however, often contains less water vapour and is a much deeper blue.*

poor clarity at night. Britain, for example, is not generally thought of as having a humid climate, but humid does not necessarily mean warm. It can be cold and humid – as in a November fog, when the air is saturated with water and the relative humidity is 100 percent.

In the daytime the difference that water vapour makes to the atmosphere is quite evident. You can generally tell by the colour of the sky, or by the quality of the air itself as viewed against distant objects, what is contributing to the haziness. Water vapour turns the sky milky. When it is very humid the sky is pale blue even at the zenith, and near the horizon the sky is virtually white. This is quite different to the effect of industrial pollution, which often lends the sky a reddish or brownish tinge, and to heat haze, caused by dry particles, which is a bluish grey.

Water vapour is as much of a problem to the urban astronomer as streetlights are. If the skies were crystal clear, free from all water vapour and dirt, those faint objects that we strain to see would be much easier to observe. It is because the water vapour in the atmosphere reflects light shining up from the ground back down again that the night sky takes on its orange or pink tinge. Not only does the water vapour reflect artificial light in this way, it also dims the light from the distant objects we want to observe. The two effects combine to prevent us from seeing anything but bright stars.

The amount of water vapour in the atmosphere varies considerably from place to place. In Britain and along the US East Coast, for example, the sky is usually milky. Very rarely do observers there get to enjoy the deep blue skies characteristic of such places as Australia and Arizona. As Figure 2.1 shows, even near Australia's capital city the skies are usually much bluer than British skies. An Australian woman who visited Britain for the first time commented that the sky seemed "much closer than in Australia." Perhaps its milky appearance made it seem nearer, like clouds, than the distant deep blue she was used to.

The reason why Britain in particular has milky skies is its geographical location. Not only is Britain an island at the edge of the Atlantic, it is also right in the path of atmospheric *depressions*, or *troughs* – regions of low pressure – which trundle across the ocean, picking up moisture as they go. As tennis matches at Wimbledon are interrupted by rain yet again, the citizens of Paris, a little over 300 km (200 miles) to the south, may be enjoying unbroken sunshine. The tracks of the depressions vary from year to year, sometimes moving farther south so that Paris shares the same grey skies. In other years the depressions favour Iceland, and Britain enjoys a hot summer.

Such weather systems are by no means restricted to Britain and the US East Coast: they affect most continental landmasses. Yet a depression is not necessarily bad news for the observer. It may well bring rain, but a passing weather system can bring clear skies as well – really clear skies, not those hazy ones that are so bad for urban astronomy.

A depression forms at temperate latitudes when a bubble of warm, moist tropical air pushes into an area of cold, dry polar air, as depicted in Figure 2.2. The air pressure at the edge of the bubble is low, and the bubble forms into a spike with its pointed end in the centre of the low-pressure area. The winds within a depression always blow anticlockwise in the northern hemisphere, clockwise in the southern hemisphere. The leading edge of the spike of warm air is known as a *warm front*, and the leading edge of the cold air as a *cold front*. On a weather map, fronts are indicated by the familiar semicircular and triangular symbols.

Clouds form along fronts, and the passage of a front, either warm or cold, is often marked by rain. After a cold front has passed by, the weather turns fresher and brighter as the cold, drier polar air washes over the land. This polar air is usually a great deal clearer than the warm humid air which preceded it. The skies take on a bluer appearance by day, and are

Figure 2.2 *A frontal system begins where mild temperate air meets cold polar air. The two air masses have different densities, so they do not mix but instead slide past each other. A ripple in the layer between them can develop into a larger feature, which eventually becomes a fully fledged depression. The triangular symbols indicate a cold front, the semicircles a warm one.*

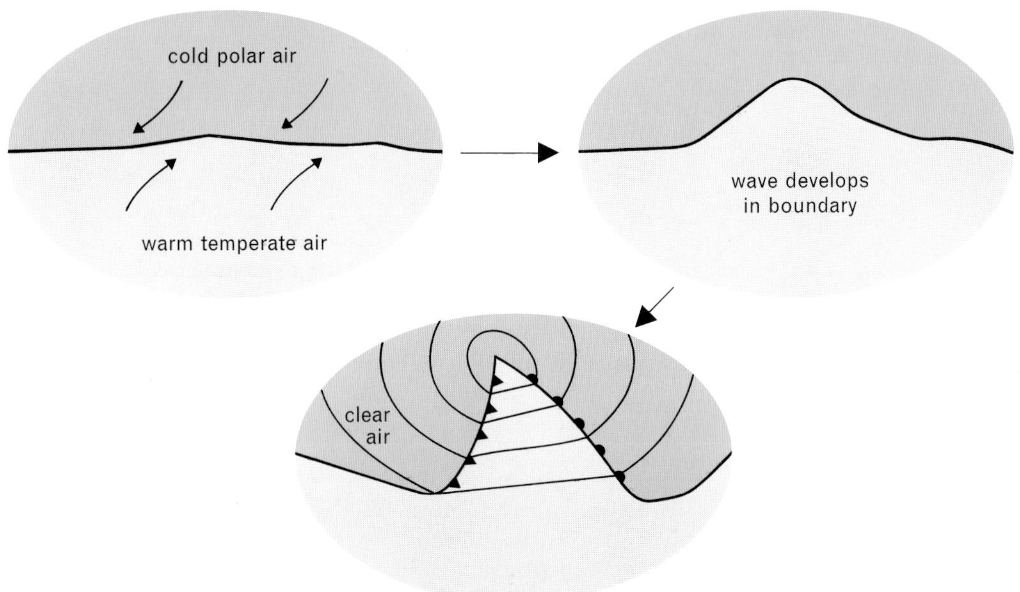

much darker by night. The haze has gone. You will then hear recited Myth Number 2: "The rain has washed the air clean."

Although it looks that way, that is not really what has happened. No doubt the rain has washed any dirt particles out of the air mass that was over you at the time, but, as long as the wind is blowing, the air that surrounds you at midday is miles away by 1 p.m. Clear skies can follow rain because they belong to a different air mass which has moved in to replace the one that was there before.

To the city astronomer this clear air is just the ticket. For a few hours at least there is the chance to observe those faint objects that have been eluding you. Your main problem may well be some cumulus clouds scudding along, making observing rather frustrating. Even so, some of the best views of the sky can be had in gale-force winds. In between the clouds the skies are very transparent, with almost no suspended dust or water vapour to reflect the streetlights. Observer and telescope must be well protected from the elements, both for the sake of comfort and to prevent the telescope from wind-induced vibration. The scale of this vibration is small, but if you are photographing a celestial object it is crucial. A telephoto lens is heavy and exerts a considerable leverage on the camera mounting, which can vibrate imperceptibly. You will probably not suspect that anything is wrong until your photographs come out unaccountably blurred.

Making the most of the occasions on which a cold front has just passed through is an important part of beating the streetlights. Unfortunately, such opportunities can be rather rare, though it does depend very much on where you live. Not only must the front have passed through at the right time, but the skies behind it must be free from cloud – rarely do weather fronts behave as they do in textbooks. And, of course, you must be out there to make the most of it.

In some areas, winter means snow. While cold air streams can be low in water vapour, snow will have an adverse effect on sky brightness – light from streetlights which would otherwise be absorbed by the dark ground is instead reflected up into the sky with high efficiency. This can counteract the advantages of the clear air. If there happens to be an aurora as well, you can probably forget deep-sky observing!

When water vapour is welcome

There are actually occasions when water vapour is welcomed by astronomers. Myth Number 3 states that "Mist creates good seeing." By *seeing*, astronomers mean the steadiness of the air through which they are observing (though sometimes the term is wrongly applied to atmospheric clarity, or *transparency*). The state of the seeing is particularly evident when you are looking at the Moon or planets. Only too often the limb (edge) of the Moon, for example, seems to be in constant motion. It shimmers and shakes as if the whole lunar surface is in turmoil. Your view of a lunar feature is spoiled as the image dances around, blurring, growing, shrinking, doubling. Details on a planet such as Jupiter are frustratingly difficult to pick out; just as soon as you see them they are gone again. The annoying thing is that the night looked so invitingly clear, with the stars twinkling beautifully. When the seeing is good the planets are rock steady, their images showing hardly a movement.

A widely used practical scale of seeing was devised by the French astronomer Eugène Antoniadi, and is named after him. It runs from I, denoting absolutely perfect, without a quiver, to V, for terrible. As far as I am aware, there is no conversion between this subjective scale, used largely by amateurs for planetary observation, and the objective scale of seeing for point sources observed with large professional instruments, which is derived from image size and measured in arc seconds.

Professional astronomers have searched the world for sites where the seeing is as close to perfect as possible, the goal being 1 arc second or below. At a small number of sites, seeing as good as 0.25 arc seconds has been achieved. (The seeing does also depend on whether you use a small or a large telescope.)

Although good seeing is particularly important when observing details on the Moon and planets, it is much less of a factor when looking for deep-sky objects: good transparency is then what really matters. However, when searching for difficult, faint objects, such as remote galaxies or the faintest stars, poor seeing can still be a hindrance.

Generally speaking, however, the two sorts of observing are mutually exclusive. Clear, sparkling nights are often accompanied by bad seeing, while nights when the planets beam steadily down without so much as a shiver are often rather misty. But back to the myth that mist improves the seeing. What is really happening is that still air, which gives good seeing, also encourages mist to form as the ground radiates its heat away into space and water vapour condenses out of the air, forming dew where it comes into contact with cold surfaces. (The exposed lenses and mirrors of your telescope are ideal surfaces for the formation of dew – another problem for the amateur astronomer.) So, although planetary observers often regard a little mist with benevolence, it is really just a sign that the seeing is likely to be good, rather than the cause of good seeing, though the water vapour does help to even out temperature differences.

The steady air which gives rise to good seeing is most likely to come during a spell of high atmospheric pressure – sometimes called a *ridge*. The daytime skies may become hazy, and the wind may drop. Sometimes, especially in winter, these conditions can also be accompanied by a dreary blanket of cloud, or by bitterly cold, frosty weather. Mel Bartels of Eugene, Oregon, describes how conditions in his area can change:

> Local winds do little harm to the image, though they shake poorly mounted 'scopes. High-level winds destroy the image when a front (usually a cold front) first passes through. A typical progression here in the NW goes like this: cloudy for a while, then just after a front passes through, very transparent but with poor seeing, then the following night, good transparency and good seeing, followed by a night or two of gradual transparency loss, then come the clouds to start the cycle over again. In the last half of summer and first half of fall, we typically have clear skies for just about the entire time, with transparency and seeing at their best for the first several nights after things clear …

This progression, however, is rarely experienced by British observers, who often have to put up with cloud, followed by rain, followed by cloud again.

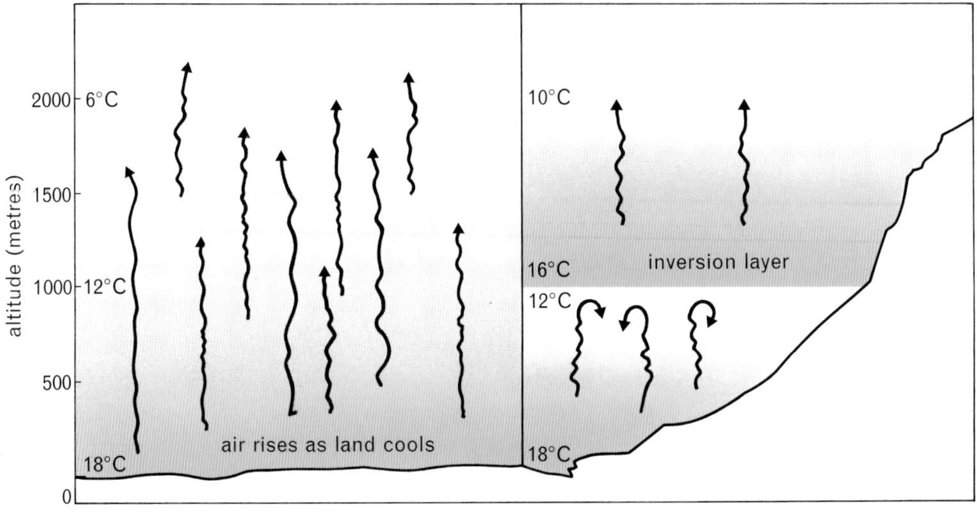

Figure 2.3 *Normally, air temperature decreases with altitude at an average rate of 0.6°C per 100 metres (1°F per 300 feet), as on the left. But where there is a temperature inversion, warm air overlies colder air, and over a small altitude range there is an increase of temperature with height. The pollution-laden air near the surface is trapped below the inversion layer, above which the air is stable and clear. Inversion layers usually occur at heights of 1.5 km (1 mile) or more, but occasionally they reach almost to ground level.*

It therefore pays to keep a close eye on the evening weather forecast. Unfortunately, some TV forecasts avoid showing a proper weather map with standard symbols, or even giving any indication of what is really happening. You may have to work out what is going on from the presenter's chatter.

Figure 2.4 *This view from the Pic du Midi, at an altitude of 2800 m (9200 ft) in the French Pyrenees, shows a temperature inversion marked by a flat layer of haze. The telescope dome in the foreground is set apart from the main observatory building in order to avoid local seeing effects caused by the concrete structure.*

Good seeing is also found above what is termed an *inversion layer* (see Figure 2.3). The normal temperature profile for the atmosphere is that it gets cooler with increasing altitude, and on a large scale this always applies. But if the lower levels are cooled by some other agency, such as a cold sea current, the reverse may occur. This is what happens on the eastern side of the Atlantic and Pacific Oceans: the west coasts of Europe, Africa, and the Americas are washed by currents which are unusually cool, having brought water from polar regions. On a small scale the normal temperature profile of the atmosphere is inverted, a layer of cold air forming beneath a warm layer. The two layers are of different densities, and stay as separate as oil and water. The air in the lower layer is unable to rise, causing the notorious fogs and pollution of such areas as San Francisco.

Above this layer, however, the air is both stable and clear. Temperature inversions are not to be found everywhere, and they are normally some way above sea level, so observers in London or New York are unlikely to benefit from them. But if you are fortunate enough to encounter one, you will find conditions ideal for astronomy. Figure 2.4 shows an inversion layer as seen from the Pic du Midi Observatory in the French Pyrenees.

The search for good seeing

Whether you live in the town or the country, what the seeing is like will determine how successfully you can observe the planets. So the question is, does the town give any worse seeing conditions than the country?

First, there is high-level seeing and local seeing. *High-level seeing* is determined by weather systems, such as the passage of fronts and the presence of jet streams (high-altitiude winds), and affects both urban and rural areas equally. You can often see the turbulence that causes poor high-level seeing by defocusing the telescope while looking at a bright star. Turn the focuser so that the eyepiece comes out towards your eye, and you are focusing on a point closer than the star. The star will expand to a disk, against which you can see what appears to be a rapid flow of material, rather like looking at the surface of a turbulent stream. You are actually focusing on a point only a few hundred metres from the telescope, though the disturbances you see may be farther away than that.

In the early 1970s I was stationed for a while at the Pic du Midi. Astronomers there experimented with focusing on this turbulent layer in order to estimate its height. Unfortunately it proved too difficult to focus on accurately, even though they were using telescopes with very long focal length, giving much finer discrimination of focus position.

Local seeing is caused by disturbances in and around the telescope itself. These begin within the tube, more so with reflectors, which are usually open at the top end, than with reflectors, in which the air in the tube is sealed in. People often complain of bad seeing when in fact they have simply not allowed the telescope sufficient time to cool down. If the air in the tube has not cooled to the temperature of the night air, turbulent currents will be set up, which will ruin the seeing. Ways of improving your telescope are dealt with in more detail in Chapter 5.

Even refractors are not immune from this. I have seen an expensive refractor take at least an hour to settle down after having been brought out from the warm. The urban astronomer's telescope may well be kept indoors, so this is an additional source of problems. The solution is to put the telescope outside a good hour or so before observing, though with all the covers in place to prevent its optics from dewing up.

Moving outwards, the next common cause of bad seeing is air currents generated by heat from the observer. This happens particularly with reflectors, where the observer peering into the eyepiece is very close to the telescope's line of sight. The remedy is simple – position yourself so that you are not beneath the tube, and make sure you don't breathe into the line of sight. The ideal observer would not breathe at all.

This effect is worse if you have an observatory. It is natural to invite people round to look through the telescope. Those who are not observing usually peer out of the dome opening, right under the top of the tube. The heat generated by each person pours out of the opening, giving lousy seeing which miraculously improves as soon as they have all gone home.

Beyond your telescope there are many other causes of bad local seeing. Anything which collects heat during the day is a potential problem. Not only its colour but its thermal properties are significant. Dark objects collect more heat than light ones, while good insulators (such as concrete) will lose heat more slowly than good conductors (such as metal). The city is full of concrete, and holds its heat well. Temperatures are usually a few degrees warmer in the centre of town compared with even the suburbs.

There are some obvious features to avoid. A neighbour's chimney is the first thing that comes to mind, but these days fewer people light fires and many chimneys are unused. Homes are warmed instead by central heating, which is much more sneaky. The vents, along with those of air conditioning units, are generally on a side wall, and can shoot a jet of hot, moist air several feet out from the building, just where you least expect it. The boiler probably switches on and off every so often, leaving you puzzled about what is going on. If the boiler provides hot water, it can be operating even on warm summer evenings.

Uninsulated roofs and tarred driveways are notorious harbourers of heat. Many city observers find that the seeing deteriorates if the object they are viewing moves over a roof. Grassy areas, on the other hand, are pretty innocuous, and an expanse of water can be ideal. But country-dwellers need not feel completely smug: broadleaved trees, when in leaf, have been accused of releasing parcels of heat throughout the night. Taking this into account, you might try moving your telescope around, if it is portable enough, in order to get better seeing. It is possible that you can get an improvement in seeing simply by avoiding a local source of stored heat.

At La Palma in the Canaries, the roads that pass by the clutch of observatory domes there have been covered with fine white gravel to prevent them from heating up unduly during the day, and giving off heat at night (see Figure 2.5). At other observatories, conifers are planted around the domes. They keep the ground cool, and – unlike broadleaved trees – are said to release relatively quickly any heat they have stored from the daytime. Observatory domes are usually painted white to keep them cool during the day. The aim is to maintain their interiors at nighttime temperatures so that no time is wasted cooling everything down after opening the dome.

It may sound as if observing from the town is pointless, with all the sources of bad seeing around. Many people will tell you that city seeing is

Figure 2.5 *The approach road to the Isaac Newton Telescope on the island of La Palma in the Canary Islands has been covered with white gravel, visible at the lower left in this photograph, in an attempt to eliminate the effects on seeing of the dark tarmac, which at night gives off heat it has absorbed during the day.*

terrible, even without having experienced it. Yet urban observers often manage to make perfectly good observations. Planetary observer John Murray used to live in Luton, Bedfordshire, just a short distance from where the legendary telescope maker Horace Dall had his observatory overlooking the urban sprawl. John found that the seeing when viewing over the town could be excellent. His line of sight was a hundred metres or so over the rooftops, so that individual local sources of bad seeing were evened out. Others who have observed from both town and country tell similar stories and confirm that in practice they notice little difference when viewing the planets.

David Frydman has observed from built-up areas in London and Helsinki for many years. He thinks that city seeing is actually better than in the country. As long as the line of sight is well above obvious disturbing influences, it can be exceptionally good. He also finds good seeing when the nighttime and daytime temperatures are very similar, and as a consequence there is little escape of heat from the ground. In Helsinki this happens in the autumn, when the sea temperature of the nearby Baltic and the air temperature are very similar, about 16°C (61°F). Mel Bartels confirms this from his own experience in Oregon.

Although the planets are bright enough to cut through murky skies, your eyes do need a certain brightness level in order to make out low-contrast detail. Jupiter is bright enough for you to make successful observations, even through misty skies. It is an ideal object for observing from the city. Saturn, however, is dimmer and has lower-contrast features. Poor transparency can therefore wipe out the features you are looking for, even if the seeing is perfect.

If your taste is for deep-sky objects, however, David Frydman recommends that you concentrate your efforts on objects that are above about 50° altitude. Below that level there is a marked cutoff point as the transparency of the atmosphere decreases and the brightness of the sky increases, as Figure 2.6 shows. I suspect that many city-dwelling astronomers give

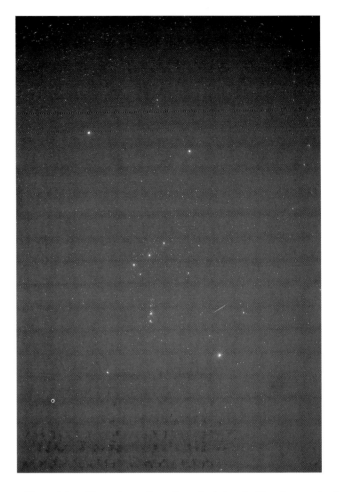

Figure 2.6 *The visibility of faint objects depends on their altitude. This photograph of Orion shows that above about 45° – in this case, the altitude of Betelgeuse – the sky becomes considerably darker.*

up deep-sky observing simply because they have become disillusioned by looking for objects when they are too low in the sky.

Seeing can improve during the night, as temperatures settle down and the ground loses its heat. The dedicated observer who prefers to view the morning rather than the evening sky can be rewarded with much better conditions. Optical worker Jim Hysom remembers the perfect seeing he encountered while serving with the British Army in Gibraltar. He was on guard duty, equipped with a telescope. He found that just before dawn he could obtain pin-sharp images of the Algerian coastline, about 20 km (12 miles) away. But a few minutes before sunrise the seeing would suddenly change to the familiar daytime hazy blur. The cause of this, he believes, is that infrared radiation from the Sun, which is refracted more by the atmosphere than its visible light output, "rises" first and warms the air before the Sun itself rises.

Solar observers have more to contend with than their nighttime counterparts, since the source of the heat that spoils the seeing is necessarily present. Local seeing is likely to have a greater effect than at night, since surfaces are being heated all the time, even on a cold winter's day. Geoff Elston, of the British Astronomical Association's Solar Section, says that even from his observing site in Clapham, south London, the seeing for solar observing can be good, mostly in the earlier part of the day.

At the Pic du Midi they have tried many tricks to get better seeing. One solar observatory dome, known as the Teapot, has a tube jutting out from it instead of the usual opening section. This intercepts the Sun's light a short distance from the dome, avoiding any turbulence around the dome's surface. Big Bear Solar Observatory in California is set in the middle of a lake, which steadies the local seeing. At other solar observatories, mirrors high above the ground collect the Sun's light and beam it down to where it is needed.

Interestingly, the seeing also depends on your telescope's aperture. The French astronomer Audouin Dollfus, who has observed extensively from a wide variety of sites (including high-altitude balloons), has offered the following explanation. Particularly at night, when the ground is cooling and the air above may be warmer, the lower atmosphere consists of layers of air of different temperatures. As air is a poor conductor of heat, these layers tend to remain separate, with well-defined boundaries between them. The layers and their boundaries undulate, perhaps even developing eddies like the swirls in a flowing stream.

As a result, the incoming wavefront of light from a star, which should in theory be smooth, is distorted on a scale which varies from 0.1 to 1 metre (4 to 40 inches). These deviations in the wavefront are often termed *cells* of air. (They are not necessarily the same as the air pockets and other forms of turbulence encountered by the pilots of light aircraft.) The beam of light from a star to your eye is being distorted by these cells, which are constantly in motion. It is this movement which gives rise to twinkling. You can get a quick idea of how good the seeing is by assessing the amount of twinkling: the less twinkling, the better the seeing.

When night and day temperatures are similar, the whole lower atmosphere tends to settle down and the layers disappear. This is what happens where there is a ridge of high pressure. Air saturated with moisture (giving a relative humidity of 100 percent), as it is when there is a mist, conducts heat more easily and will thus lose any layered structure it may have had. As mentioned on page 15, the seeing improves under such circumstances.

Amateur astronomers often say they get better seeing with a small telescope than with a large one. The reason now becomes clear: if the seeing cells are about 100 mm across, then a telescope of roughly this aperture will be looking through little more than one cell at a time. The image will be sharp, and although it may move around a bit, it will still be easy to observe. A larger telescope at the same site will be looking through two or more cells at the same time, with inevitable blurring. This is why you sometimes see double or triple images. I suspect that this effect is one of the reasons for the claimed superiority of refractors over reflectors, for refractors are generally of smaller aperture.

If observing from the city is deemed to be a waste of time, then observing from a high-rise building or an upper-storey window is surely total folly. There are several factors which conspire to ruin your view. First, the building will be giving off heat it absorbed in the daytime, setting up an invisible wall of bad seeing around it. Second, observing from an open window means letting all the heat in the room flow out of the window, passing around the telescope, which is also bound to ruin the seeing. And third, you just cannot get a steady enough base: an upper floor of a building is far too susceptible to vibrations compared with terra firma.

That, at least, is what some people say, but others have proved them wrong. The balcony of a high-rise may not be the best place to observe from, but it does have its advantages. Not the least of these is the fact that it is accessible. I have always said that the 150-mm (6-inch) telescope set up at home is worth far more than the 400-mm (16-inch) half-an-hour's drive away. If you can get ready within minutes, you are much more likely to observe than if you have to spend ages getting everything into place. If the building has concrete floors, you should have a steady enough base. Your main worry will probably be the risk of your precious equipment toppling over the balcony.

Although observing through an open window is notoriously bad for the seeing, all amateur astronomers have done this at some time or other, and the results are not always as bad as some would have you believe. It is even possible to carry out useful work from indoors, as detailed in Chapter 8. And if your options are to observe through a window, or go outside and risk being mugged, I know which I'd choose.

One big problem with a high-rise is likely to be turbulence around the building. This, however, will vary with the wind direction. With experience you may find that some winds are more favourable than others. It is probably better to be facing into the wind than in the lee of the building, though the effects may be masked by the fact that different air masses come from different directions anyway. North winds, for example, usually bring different weather from south winds, wherever you live. In fact, it is possible that under the right circumstances the seeing from a high-rise could be very good, since you are above local sources of bad seeing. Professional astronomers, too, have discovered that good seeing is to be had high above the local ground level, and accordingly locate their telescopes well above the ground. The aim is to be in a smooth air flow, with no local eddies.

Even mountaintop observatories have problems with turbulence. At the Pic du Midi, for example, the mountain is the first really high ground that northerly winds encounter. The air mass rushes up the slope and swirls around right over the observatory, creating bad seeing. Winds from the southeast round to the west, however, meet the higher mountains of the main Pyrenees chain first. They dump their precipitation on the mountains, leaving a comparatively tame air mass to encounter the Pic itself, which is in their lee (though a long way downwind).

Such effects are not restricted to major mountain chains. Many cities are in the "rain shadow" of a range of hills, and it is even possible that city buildings can act as a barrier, if not to rain then at least to turbulence, in the same way.

How bad is the weather?

There are some lucky amateur astronomers who live in places where the skies are often clear and the seeing is often good. But the rest of us have to put up with weather that varies from the mediocre to the atrocious. British astronomers feel particularly hard done by, with clear skies between only 20 and 33 percent of the time. Even when it is clear, the weather seems to know just when you are about to observe. No sooner have you got your coat on and set up the telescope than cloud appears just where you were going to look. You wait a while, then give up. Ten minutes after you pack up, the sky is clear again.

Despite this, some great astronomical discoveries have been made from Britain. British amateurs today have to put up with the same weather as, for example, did William Herschel in the eighteenth century. During one night's observing, it became so cold that the ink froze in his inkwell. Admittedly he did not have light pollution to contend with as well, but my point is that there is enough clear weather even in the worst climate to allow the keen amateur some observing time. The main problem is flexibility. In a poor climate you cannot reserve Thursdays for astronomy, or decide to have a Saturday night star party – you have to grab any opportunity when it presents itself. The difficulties are really social rather than astronomical. So Myth Number 4, that in an unreliable climate it is not worth observing, simply reduces to a question of priorities.

By getting to know the seasonal weather patterns in your area, you can begin to overcome some of its worst effects. Keep an eye on the weather forecasts, and learn what sort of weather gives you the best conditions. You may have to adapt your observing methods to the weather. Rather than wait for perfect conditions for wide-field deep-sky photography, which requires continuous clear skies, for example, try planetary photography instead, or aim for star clusters that can be recorded with shorter exposures through the gaps in the cloud. Either that, or buy a CCD (see Chapter 6).

CHAPTER 3

Know your enemy – the streetlights

When I was young – and I'm not all that old – the bane of my astronomical life was the streetlight across the road from my back garden in suburban Middlesex. It had a tungsten bulb of maybe 100 watts, and went out at about midnight. I remember looking out of the window at about 3 a.m. to see the dark skies, though my enthusiasm did not stretch to going outside to observe. The evening skies were then still tolerably dark, allowing me to see plenty of objects.

I also remember a holiday trip to the small seaside resort of Dawlish Warren in Devon at about the same time, in the early 1960s. One night it was cloudy, but I thought there might be some clear patches. I made my way down to the shore, but as soon as I moved away from the lit footpath it became so dark that I could not find my way. Today there would be no difficulty in walking around on a cloudy night – the glare from nearby Exeter and other towns, reflected from the clouds, would provide ample illumination. Streetlighting, and other forms of outdoor lighting, have spread far and wide over the past 30 years.

Streetlighting began longer ago than might be imagined. As early as 1405, the Aldermen of the City of London had to ensure that every house alongside a road displayed a lit candle in the window from dusk until 9 p.m. Four centuries later, one of the first uses of town gas (a byproduct of coal, now superseded by natural gas) was to provide streetlighting. In 1765 a colliery manager by the name of Spedding suggested that the streets of nearby Whitehaven in the north of England should be illuminated by gas lights similar to those he had installed in his offices, which used gas from the mine. His idea was turned down. In the end, it was the streets of London that were illuminated by the first public gas lighting, starting in 1807 when lights were installed in Pall Mall by Frederick Winsor.

From then on, gas lighting spread wherever gas companies were set up. Towns became brighter. In his famous novel *Far From the Madding Crowd*, written in 1873, Thomas Hardy refers to the glow from the nearby town of Casterbridge, which in reality was Dorchester, about 5 km (3 miles) from the cottage where Hardy spent his early years:

> A heavy unbroken crust of cloud stretched across the sky, shutting out every speck of heaven; and a distant halo which hung over the town of Casterbridge was visible against the black concave, the luminosity appearing the brighter by its great contrast with the circumscribing darkness.

When within a mile or so of the town, Hardy's character Fanny Robin can see her way "by the aid of the Casterbridge aurora ..."

Such lighting must have been much dimmer than present-day urban lighting. Its original aim was to help citizens find their way around without walking into potholes, puddles, and other undesirable features of the thoroughfares and byways of the time. A gas lamp every hundred metres or so was adequate for this purpose. These days, most of us rarely find ourselves out on a truly dark, cloudy night where there is not a glimmer of light from earth or sky. Few non-astronomers keep track of the phases of the Moon, for example, yet to our forebears they were important. In the eighteenth century a group of eminent men calling themselves the Lunar Society met regularly in Birmingham. They chose the name not because they were particularly keen on the Moon, but because they met each month around the time when it was full, making it easier to get to the meetings.

Eventually, gas streetlights were replaced by electric ones, generally of similar brightness to the gas lamps. Some gas lamps remain in London, usually in touristy areas where they are kept for their picturesque qualities. But some can be found even in the side-streets of the West End. Members of the Royal Astronomical Society or the British Astronomical Association emerging after a winter meeting may not notice that the small street outside the entrance to the Scientific Societies' Lecture Theatre is actually lit by gas lamps. In fact, most of the light in the street streams through the windows of the buildings which line it.

During the Second World War, blackouts were strictly enforced in Britain and other countries to avoid giving clues to enemy bombers. The population again had to learn how the Moon's phases vary, in some cases because its light meant that an air raid was more than likely. Many Londoners remember seeing the Milky Way for the first time during the blackout, and quite a few became interested in astronomy as a result. In California, Walter Baade found that he could push the famous 100-inch (2.5-metre) telescope on Mount Wilson, overlooking Los Angeles, to its limits. Since the end of the war, however, streetlighting has increased in brilliance virtually everywhere. Astronomers wish the blackout could return – but without the attendant risk of air raids.

Types of streetlight

The most familiar type of electric light, the sort most of us use in our rooms, is the tungsten bulb. It consists simply of a wire filament inside a glass vessel from which all the oxygen has been extracted and replaced with an inert gas. The gas plays no part in the light-producing process, but prevents the filament from evaporating as the current passes through it, heating it to incandescence. Although we use the white light it gives off to see by, most of the energy – up to 98 percent – is converted to heat rather than light. This really is wasteful, and it is for this reason that energy-saving light bulbs and tubes have come onto the market.

A variant on the tungsten bulb is the tungsten halogen lamp, in which a halogen gas such as iodine is added to the inert gas in the bulb. This prolongs the life of the filament and allows it to be run at higher temperatures, giving a whiter light.

Studying the spectrum

An important feature of any light source is the spectrum of the light it emits – that is, the distribution of its light across various wavelengths. For astronomers it is crucial to know the spectrum for each source of unwanted light so that some counteractive measure can be taken.

Looking at the spectrum of white light, as produced by a prism or diffraction grating, for example, we see the familiar rainbow of colour. What we do not see is that the rainbow extends on either side into "invisible" radiation. Beyond the red end is infrared (which we can feel as heat), and beyond the blue end is ultraviolet (which tans our bodies and can have other, more serious effects on our health if we get too much exposure to it). The word "radiation" covers anything which radiates, from plain old yellow light and ultraviolet, which are both types of electromagnetic radiation, to particles such as those given off by radioactive substances.

What we see as white light is really the visual impression of all the indvidual colours seen together. To be more accurate, white light is radiation from a body at a particular temperature – for example the Sun, whose visible surface is at a temperature of around 5500°C (10,000°F). A block of metal heated to this temperature would give out virtually the same proportions of colours – there are additional absorptions in the solar spectrum – and the same amounts of infrared, ultraviolet, and other forms of radiation.

The feature that distinguishes the different types of electromagnetic radiation is their wavelength. Traditionally, wavelength was measured in angstroms (symbol Å), on which scale the wavelength of yellow light is around 5000 Å. Today, it is measured in nanometres (nm), 1 nanometre being 10 angstroms. So yellow light has a wavelength of 500 nm, or 500 billionths of a metre; red light has a wavelength of around 650 nm, and blue light around 400 nm. With sunlight, starlight, or a tungsten lamp, the light consists of all colours, though there is rather less blue and red light than there is yellow. They have what is called a *continuous spectrum*. In Chapter 6, I deal with ways of filtering out the glare from streetlights, but since the spectrum of a tungsten lamp is virtually the same as that of a star, tungsten lights are a problem. Overall, they contribute little to the upward-directed light pollution that we have to deal with, though on a local level they can be very irritating in the form of security lights (now often disparagingly referred to by astronomers as "insecurity lights").

Mercury lighting

Although many small towns were once lit by tungsten lamps, like the one shown in Figure 3.1, these lamps were found to be very wasteful when brighter lighting came to be needed. Lighting engineers therefore turned to other sources of light. An early success was the mercury vapour lamp, in which a high electric current is passed through mercury vapour, which then gives off an intense bluish-white light. The result is a bright lamp, well suited to lighting streets. In the late 1950s the old tungsten lamps in the road alongside my garden were converted to mercury lamps. Fortunately, they still had a time-switch which turned them off at midnight. But I soon found that mercury lighting destroys night vision.

This is because the night vision of the human eye is controlled by a sequence of chemical reactions which alters its sensitivity to light. At night,

Figure 3.1 *A picturesque streetlight from the early twentieth century. It has a simple tungsten bulb and uses mirror segments to reflect the light sideways so as to give a wider spread.*

your eyes are most sensitive to green and blue light, as emitted by mercury lights, so one glance at a bright green light is enough to ruin your night vision for quite a while, until the chemical reactions have had time to restore it. Although the iris of your eye contracts and expands as well, this has only a very small effect on the overall sensitivity of the eye. The iris provides a short-term means of controlling the light received over a narrow range. Our eyes take time to become dark adapted when we go out from a brightly lit room into the night. The iris expands almost immediately, but the chemical reactions that give us our most sensitive vision take much longer. You should allow at least half an hour to get dark adapted; it is likely that your night vision will improve even after this.

The limited control provided by the iris will be appreciated by anyone who has their eyes tested by an ophthalmologist and is given eye-drops to dilate (expand) the pupils. When fully dilated the pupils may widen from about 3 mm in artificial light to almost 10 mm. Even when they are this wide, however, it is possible to go outside into bright sunshine with no more than a pair of sunglasses to make the light bearable. This experience shows that the iris has only a small effect on the eye's sensitivity: the chemical reactions are much more important.

Until recently the French used yellow headlights for their cars rather than white ones, partly in order to reduce glare. It is certainly true that facing an oncoming stream of yellow headlights is much preferable. Now the French have had to adopt the same standards as the rest of Europe, and are switching to white headlights. Yellow night-driving glasses to help reduce headlight glare have recently come into vogue. I find that while I dislike driving at night wearing yellow glasses, they work just as well as

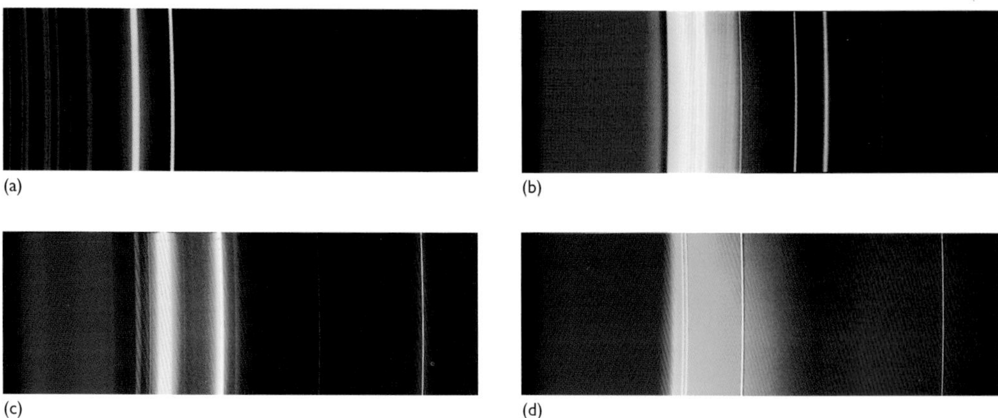

Figure 3.2 *Spectra of streetlights. (a) Low-pressure sodium, with a dominant pair of orange-yellow lines, here shown as one. The green line is in fact very weak and contributes very little to the light of the lamp. (b) High-pressure sodium, with a dark band at the main wavelength of low-pressure sodium. (c) Mercury, with enhanced red emission as a result of coating the lamp with phosphors. (d) Fluorescent, with additional phosphor coating inside the tube to increase the blue and red output. These photographs of spectra show more detail at the red end than can be seen by eye, and the colour rendering of the photographic emulsion produces fewer individual colours than are visible with the eye.*

ordinary sunglasses when I am reading in bright sunlight, since they cut out the blue light that is so painful.

It is because blue light destroys night vision that astronomers making notes of observations at the telescope habitually use red light, as did the usherettes who used to show you to your seat in the cinema (now you have to stumble there by yourself). Red light has very little effect on your night vision, and you can carry on observing almost immediately after recording your observation.

Mercury streetlights have a different type of spectrum from tungsten bulbs. Instead of a continuous spectrum, the light is concentrated into very narrow lines of single colours – an *emission line spectrum*. There is a strong blue line, a green one, a yellow one, and a rather faint red one. This means that it is possible to see colours under mercury lighting. For you to see a particular colour in an object, it needs to be illuminated by light that contains that colour in the first place. A red car appears black if there is no red component in the light that illuminates it, for example. As the red line in mercury lighting is weak, reds are rather hard to see by it. The spectra of different kinds of streetlight are shown in Figure 3.2.

These days, mercury streetlights are fairly uncommon in Britain, but they are still widespread in some parts of the US and other countries. They are also used for decorative floodlighting, as with appropriate filters they can give a strong blue or green light.

If you want to look at the spectrum of a streetlight for yourself but lack the necessary equipment, you can do so if you own a compact disc. A CD is in effect a circular diffraction grating, even though its surface bears lines of tiny pits rather than grooves, and will produce a very satisfactory spectrum. Examine a streetlight by looking at its reflection in a CD. You will see that a spectrum is produced if you look across the disc, with the central hole between you and the light source. (Actually two spectra will appear, one on either side of the light source.) As long as the source is small in comparison with the width of the spectrum, you will see any spectral lines that are present.

Fluorescent lighting

A variant on the mercury lamp is the familiar fluorescent or strip light. This starts out life as a mercury vapour lamp designed as a long tube that gives out a considerable amount of ultraviolet light. The inside of the tube is coated with a material known as a phosphor that glows brightly with a warm pinkish light when ultraviolet light strikes it. The spectrum of a fluorescent light shows a strong green line and some other fainter lines, plus a pinkish glow which covers most of the spectrum. The full range of colours can therefore be seen in fluorescent light. Curiously, the green line particularly affects the green sensitivity of colour film, which is why colour slides taken in fluorescent light usually have a strong green cast to them.

At one time it was the fashion to install fluorescent streetlights in city-centre shopping precincts since their good colour rendering made people's faces and clothing look natural. This is regarded as very important by lighting engineers – people hate lighting that gives their faces an unnatural cast. These days, however, high-pressure sodium lighting has largely taken on this role.

While fluorescent streetlighting is now comparatively rare, domestic fluorescent light fittings are still in common use, notably in kitchens. This usage contributes little to sky glow, but it is nevertheless very aggravating to the observer. If you have a fluorescent light in any room you have to pass through to get outside, you will have very poor night vision for some time. The solution is simply to switch off the light well in advance so that you do not have to look at it immediately before observing. (The rest of the household has to know that you do not want it switched on again, but this is a matter for domestic negotiation.) It may be worth putting a table lamp with a red bulb in the room so that life can continue while you observe. By the same token, it is not a good idea to watch television or use a computer immediately before going outside. The phosphors in TV tubes and monitors can be just as damaging to your night vision as a fluorescent light. Computer software intended for use at the telescope, such as planetarium programs with telescope control abilities, often has a night vision option which gives a red screen.

Other people's fluorescent lights are another matter, and the sudden appearance of a kitchen or bathroom light at a crucial moment in your observing schedule can be very unwelcome, to say the least. About the only thing you can do is to anticipate the light being switched on and choose your observing site accordingly. Fluorescent lighting is also used in offices and stores, where it may be left on all night, and on filling station forecourts. The light from these sources may spill upwards to contribute to the sky glow.

Low-pressure sodium lighting

Going back to my own personal streetlight, I was rather pleased when, sometime in the late 1960s, it was replaced with a low-pressure sodium lamp. This works in the same way as a mercury lamp, except that the gas used is sodium vapour rather than mercury vapour. Its spectrum consists of virtually just two lines of orange-yellow colour, very close together. The wavelengths of these lines coincide quite well with the maximum daytime sensitivity of the human eye, so low-pressure sodium lights provide the maximum visible light for the minimum energy consumption. This makes them the most efficient type of streetlight there is, and in the UK they are widely used on major roads. From the astronomer's point of view this is as good as we can hope for, since the twin spectral lines can be filtered out very easily, as described in Chapter 6. A low-pressure sodium streetlight is shown in Figure 3.3.

Unfortunately, nothing is ever perfect. There are drawbacks which prevent sodium from taking over from other forms of lighting. One is that it is quite impossible to see any colour other than yellow by its light; all you can see is shades of yellow. Some people intensely dislike the colour of sodium light for this reason, saying that it makes people's faces appear ghastly. It is hard to see why this should prevent it from being used on roads intended mostly for traffic. The reason often given is that it is important for the police to be able to distinguish the colours of cars, but in fact in low lighting conditions colours are hard to see anyway. And police cars, like others, are equipped with tungsten headlights, which make car colours quite easy to see. Another technical drawback with the sodium vapour lighting unit is that it is quite large. This becomes important when designing a lamp that shines only where you want it to.

Paradoxically, the sodium lamp's very efficiency can give it a bigger impact on the night sky than mercury lighting. I remember hearing an American amateur visiting Britain comment that the orange glow in the night sky, caused by the widespread use of sodium lighting, was more visually distracting than the bluish colour of mercury lighting. The eye

Figure 3.3 *Close-up of a low-pressure sodium streetlight. The lamp itself is enclosed in a glass housing which is moulded so as to refract the light sideways. This type of fitment has a considerable unwanted upward light spill.*

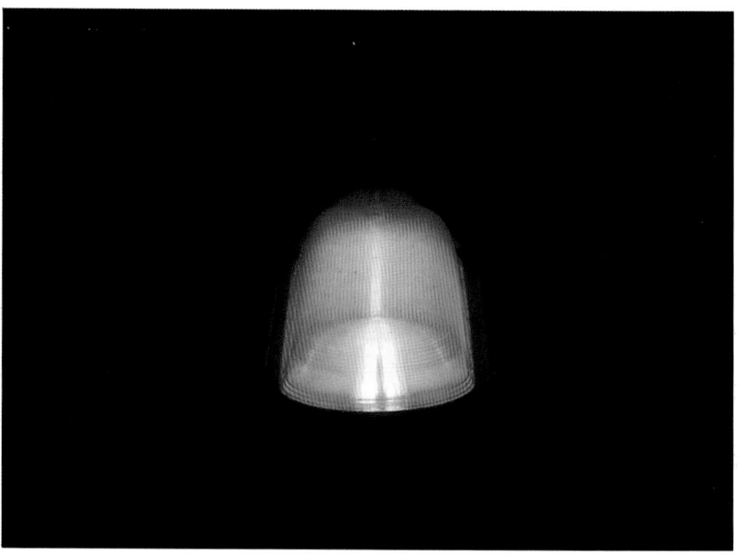

is less sensitive to blue light than to yellow, so an orange-yellow night sky looks brighter than a blue night sky.

I soon discovered one further big drawback of my local low-pressure sodium streetlight – it was fitted with a photoelectric cell, so it stayed on all night until dawn arrived. Today, time switches are virtually unheard of in Britain, and British towns are as brightly lit at 3 a.m., when most streets are deserted, as they are at 10 p.m. (Actually, many suburban streets are deserted at 10 p.m. as well.) In some areas, residential streetlights are still switched off at about midnight.

Fortunately, my local streetlight has not yet been converted to high-pressure sodium. But this type of lighting is creeping ever closer, and I suppose it is only a matter of time.

High-pressure sodium lighting

When an electric current is passed through sodium vapour at high pressure, the spectrum of the light emitted is broadened out from the two narrow orange-yellow lines into a band running from the red end of the visible region through to the blue end. Although the lights have a pinkish-yellow colour, they give an acceptably lifelike colour rendering. The lamps themselves are only two-thirds as efficient at turning electricity into light as low-pressure sodium lamps, but the installation as a whole can be more cost effective because the lamps are brighter and cover a wider area. On an eight-lane motorway, for example, fewer high-pressure lamps are needed to light the entire width.

To begin with, high-pressure sodium lighting was restricted to shopping areas – taking over the job previously done by fluorescent lighting – where a pleasing colour was considered more important than cost. But in the 1980s it started to spread out from city centres, and can now be found on many urban roads. Long stretches of Britain's motorways are now lit by high-pressure sodium lights. As more of these lights have come into use, our skies are turning pink instead of orange.

In 1993, the city of San Diego decided to replace low-pressure sodium lights, which were installed in order to protect the skies above the 200-inch (5-metre) Hale Telescope on Mount Palomar, with high-pressure sodium. It was felt that using whiter light would somehow reduce the crime rate. Cynics confidently predicted that the change in lighting would do nothing to deter criminals, and would probably make it easier for them to see what they were doing.

Like tungsten lighting, high-pressure sodium is virtually impossible to filter out. But from the astronomer's point of view there is one thing in its favour: because the light source itself is smaller than a low-pressure sodium lamp, it is easier to design a lamp housing that directs more of the light where it is needed. And herein lies what could be the answer to the amateur astronomer's prayer – the full cutoff streetlight. To see why this is important, we need to look first at the way streetlighting is used and the design of the lamp housings themselves.

Lighting a street

Streetlighting has to do different jobs in different locations. The level of lighting in a quiet suburban road, for example, does not have to be as high as on a fast major road. In city centres, lighting is as much for the

Figure 3.4 *Two different types of lighting are used on this section of the M4 motorway from London to Wales. Along the centre are high-pressure sodium lights on high columns which use a combination of reflection and refraction to create a spread of light on the road surface. At either side, along the slip roads, are low-pressure sodium lights which have a lower light output though the lamps themselves are larger.*

benefit of pedestrians shopping in the evening as it is for drivers. In some areas it may be used to make pedestrians feel safe from attack by muggers.

The main aim in lighting a street is to enable drivers to see objects that might get in their way, such as pedestrians, dogs, or other cars. Such objects do not need to be brilliantly lit, but enough light has to be put onto the road surface for them to stand out in silhouette. Brighter lighting, often giving a good colour rendering, is used on pedestrian crossings.

But there is more to lighting a road surface than putting a lamp on a post every so often. That would produce a pool of light below each lamp, but very little light in between the lamps. The farther you can make the light spread sideways, the fewer columns you need and the less it all costs. To do this, the glass housing the light source may be fashioned so as to spread the light either side of the lamp, illuminating more of the road surface, as shown in Figure 3.4. In effect, it behaves like a lens. If you look at a road surface you can often see a pattern of stripes caused by the focusing action of the housing. The lighting engineer has to make sure that the patterns from two adjacent lights overlap so as to produce as even an illumination of the road surface as possible.

It is because housings are designed to direct light sideways rather than downwards that some of it inevitably spills upwards. (With the astronomers' favourite, low-pressure sodium lighting, the comparatively large lamp can lead to considerable upward spillage.) If the spread from each streetlight were not so great, more columns would be needed, costing more money and giving drivers more things to hit at the side of the road. So there is an economic benefit in making the light spread sideways. Since sideways lighting with low-pressure sodium lamps usually produces some upward spillage, it is easier to allow a little to shine upwards than to cut it off, though with good design this can be avoided.

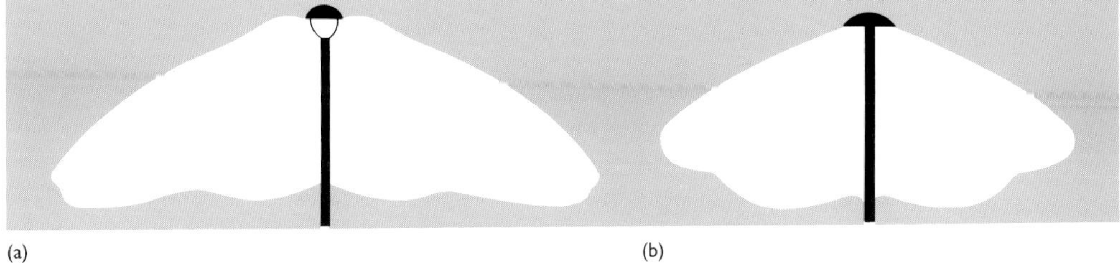

Figure 3.5 *This schematic illustration is based on polar diagrams produced by a streetlight manufacturer. Here, the extent of the white area in a particular direction from the lamp housing represents the intensity of the light as seen from that direction. Whereas the standard type of installation (a) sends some of its light upwards, the full cutoff design (b) directs all of its light towards the ground. (Based on information kindly supplied by Siemens Ltd.)*

The manufacturers of the housings publish *polar diagrams* – plots of how the intensity of light varies around the housing. To the surprise of astronomers, polar diagrams such as Figure 3.5 show only a small percentage of light directed upwards. So what is the source of the problem? We know that our night skies glow with the colour of streetlights, yet hardly any of their light is supposed to shine upwards.

One reason is that the polar diagrams show the results of laboratory measurements on brand new, clean fittings. After only a short period of use out in the open, the glass of the housing becomes pitted and dirty, causing more of the light to be scattered upwards than was intended. Another reason is that some light is reflected upwards from the road and other surfaces. Yet another is that even light directed in the first place only slightly above the horizontal can eventually find its way up into the sky. So what starts off in theory as negligible ends up in the real world as a serious light pollution problem. And it has also recently emerged that because the manufacturers have not been interested in upward shining light, they did not bother to measure it.

What is the actual percentage that shines upwards? According to figures obtained by Britain's Institution of Lighting Engineers, a typical value is 20–30 percent of the light that shines downwards from the lamp itself. A much smaller percentage is contributed by reflection from the road surface. If you fly over a city at night, you will have little doubt that most of the light shining upwards comes directly from the streetlights themselves, rather than by reflection from illuminated surfaces.

Full cutoff lighting

For some time, it looked as if astronomers would just have to get used to the sky glow, since modern life demands streetlights and streetlights cause light pollution. But progress has been made in the design of the light fittings themselves. Originally, high-pressure sodium housings were designed in the same way as those for low-pressure sodium, but new designs have made it possible to change the way in which the light is spread sideways. The beam of light is shaped by reflectors inside the housing rather than the lensing effect of the glass. There is no possibility of light escaping upwards, and the beam cuts off sharply at an angle.

These are called full cutoff or flat-glass streetlights, and the lamp itself is set inside a housing with a flat, horizontal glass window. A streetlight of this type is shown in Figure 3.6.

The benefits to both astronomers and drivers are considerable. From the same level as the lamps you can see that the beams shine downwards, and that the only light reaching the sky is reflected from the road surface. From the driver's point of view, full cutoff lighting is free of much of the glare that comes with older types. This really is a boon for driving in the rain, since much less light is scattered by droplets on the windscreen, and the driver has a better view of the road ahead.

Another group of people are happy, too. Those who live close to main roads in rural or semi-rural areas often object to the illumination of their houses and gardens by the streetlighting. With the new lamps, the spillage is much less than before. At one time these lights cost more to install, because they needed to be mounted higher up in order to give the same spread of light as before. But modern designs have made these lamps as energy efficient as the older types, even when used in quiet residential roads.

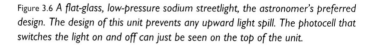

Figure 3.6 *A flat-glass, low-pressure sodium streetlight, the astronomer's preferred design. The design of this unit prevents any upward light spill. The photocell that switches the light on and off can just be seen on the top of the unit.*

(a)

(b)

Figure 3.7 *(a) A street illuminated by standard low-pressure sodium lights, with poor control of light spill. On a misty evening the upward glare from each lamp is clearly visible. Very close to where this picture was taken, the street crosses a motorway lit by full-cutoff streetlights (b), photographed under the same misty conditions as in (a). From this vantage point the lamps themselves are almost invisible, and the beam pattern shows very clearly that there is no upward light spill.*

Full cutoff streetlighting is now being installed on many sections of Britain's motorways (see Figure 3.7), and as lighting engineers become more aware of its advantages it should spread to other roads as well. In 1993, the British Department of Transport guidelines recommended for the first time that full cutoff lighting be used wherever possible. But it will be many years before even a small fraction of streetlights are full cutoff. There is a full cutoff design for low-pressure sodium, which from the astronomer's point of view has to be the ideal streetlight.

Other forms of light pollution

Streetlights are simply the most visible form of light pollution. There are many others, of varying degrees of severity. Probably the most significant are floodlights used to illuminate sports grounds, buildings, and car parks; lighting at shopping precincts, industrial plant, and filling stations; and advertising signs and security lighting. Ordinary domestic lighting usually plays only a very minor part. Less common but just as bothersome if nearby are such one-offs as laser shows. The relative contributions of these various sources to the total sky brightness varies very much from place to place.

Floodlights at sports grounds may be tungsten halogen or mercury lights, and as a rule point not downwards but at an angle towards the playing area. For this reason they can be seen over a huge distance. An increasingly common location for floodlighting in Britain is the golf driving range, which can be in operation for many hours in the evening.

Many buildings are now floodlit at night, and not just major landmarks and public buildings but also churches and office blocks. Such floodlighting is partly for prestige and partly for security, because an illuminated building is more difficult to approach at night without being seen. The cheapest way of arranging the floodlights is to put them on the ground pointing upwards – very bad news for astronomers.

A prominent feature of the night sky over the amateur observatory at Puimichel in Provence, France, is a fan of beams in the northern sky looking just like those old engravings of the multi-tailed Great Comet of

Figure 3.8 *A globe light in a parking area, a good example of an extremely poor lighting unit. The light source is mercury, which is inefficient compared with sodium, and the support obstructs some of the light shining downwards. Even though a large number of such units are used in this installation, the ground is very badly illuminated.*

1744, when the comet's head was below the horizon. The source of the beams is the floodlighting of a spectacular rock formation called Les Mées, about 8 km (5 miles) to the north. The floodlights point upwards, but a considerable part of each beam misses the rock at which it is aimed. This problem is common to most floodlighting.

Outdoor factory lighting may be there either for the security of an outdoor workforce, or to deter intruders. A short distance from my home is a depot for the London Underground railway. It is lit by huge columns topped with clusters of outward-pointing high-pressure sodium lamps that can be seen for miles.

Car parks, particularly around supermarkets, and pedestrianized shopping areas are often lit by globe lights – transparent or opaque globes with a lamp inside, often mercury. As Figure 3.8 shows, they are singularly useless at doing the job of lighting the ground since their light is totally undirected. They will illuminate any nearby buildings, which street-lighting engineers sometimes consider a good thing since it is supposed to enhance the appearance of the area. But a globe light with a mercury lamp inside it is one of the most inefficient means of lighting there is. They look pretty by day, and in the architect's drawing with all its contented citizens; but by night, glaring over an empty car park, they are a disaster. Some designs have the dubious distinction of throwing more light into the sky than onto the ground, since the base of the lamp fitting can cast a large circular shadow. This deep pool is a good place for muggers to hang out, unseen against the light from the lamp itself, but with a good view of their prey.

Because globe lights are so inefficient, they are commonly used in clusters with several units on one column. But some types of globe light are designed to throw the light downwards, using internal reflectors. Not only are these better from the astronomer's point of view, but they also illuminate the ground much more effectively.

These days, security lights are becoming increasingly common. Quite often they are mounted on the wall of a building, and arranged so as to illuminate as large an area as possible. This usually means pointing them outwards so as to deter people from approaching at night. Security lights for small businesses, homes, and car parks often have 500-watt tungsten or tungsten halogen lamps. Actually, these are very inefficient and expensive to run, and quite often are used in an ill-planned way. Their very brightness usually means that there are also pools of inky darkness, and the fact that they are on the building pointing outwards means that anyone looking towards the building who might be in a position to spot an infiltrator will instead be dazzled by the glare of the lamps.

By and large, the effects of security lights are quite localized, frustrating the individual observer rather than adding greatly to the sky brightness, since their contributions are small compared with that of streetlights in urban areas. A bright light shining in your eyes, even from miles away, is just as disconcerting as a bright sky.

David Crawford, of the International Dark-Sky Association, has assessed the main contributors to light pollution:

> We have estimated that about one-third of the sky glow problem comes from streetlights in the average community. Another third comes perhaps from sports lighting (though this can be really bad in some locations), advertising lighting, and such. The other third covers all the rest, including private lighting. It varies with the size of the community and other things. Here in Tucson, all street and parking lot lighting is full cutoff, and so the "third" must be less. We estimate that bad sports lighting and bad advertising lighting are now the dominant problems. The industrial yards are a real problem in some places, as the lighting is almost always poor. It is not too bad here in Tucson, though. Some locations, like Oakland, California, have very bad lighting at docks too.

What can be done about light pollution?

At one time, it seemed that astronomers were powerless to do anything about the encroachment of all the various sorts of lighting. Apart from a few areas near major observatories, such as in La Palma in the Canaries and Tucson, Arizona, astronomers had a very weak voice indeed. After all, they were mostly either amateurs or professionals who could presumably visit telescopes on remote, dark sites. They were very much in the minority compared with the large number of "users" of streetlights: in theory all drivers, all pedestrians, and anyone using premises lit at night – in short, most of the population, whether they were aware of it or not.

But although the problem seems to get worse by the year, there are signs that the tide is turning. In Britain and America, substantial improvements have been made by such organizations as the Campaign for Dark Skies and the International Dark-Sky Association. The aim of these groups is to raise awareness of the problem at national level, and they have been amazingly successful in doing so, considering the very limited resources available to them. At a recent lighting fair in London, for example, it turned out that a large proportion of the exhibitors were aware of the problem of light pollution, and were keen to do something about it. But despite this, there are economic arguments which weigh heavily against the astronomer.

The ways in which streetlighting can be controlled break down into control of the type and direction of the lighting, control of the time for which it is on, and control over the need for it in the first place. The circumstances governing each case vary from location to location.

Professional astronomers prefer low-pressure sodium streetlighting, since it is restricted to a very small part of the spectrum and if necessary can be filtered out. To some extent, amateurs would prefer it too, but – as British amateurs have discovered – a heavy pall of low-pressure sodium is virtually impossible to filter out. There is just so much of it, and each lamp spills light upwards. Full cutoff high-pressure sodium, however, spills so much less into the sky that UK amateurs now prefer it to low-pressure sodium. The ideal would be full cutoff low-pressure sodium, which is comparatively rare – and therefore more costly – largely through the conservatism of the lighting industry. If there were more of a demand for it, it would be cheaper.

As mentioned earlier, however, usually it is not enough simply to swap an existing glaring lamp for a full cutoff one. A completely new housing is needed. As most streetlighting installations have a lifetime of around 30 years, it would be uneconomic to replace existing fittings with full cutoff ones. So astronomers cannot hope for an instant solution to the problem of streetlight glare.

Dark skies after midnight?

The timing of streetlights these days is almost always controlled by photoelectric cells. These have great advantages from the lighting engineer's point of view. They are cheap – each switch costs only about £5 (US$8) – and more reliable than time switches. They switch on automatically whenever it gets dark, whatever the weather, and, unlike time switches, there is no need to compensate for the changing hours of daylight over the course of the year. And they do not get out of step if there is a power cut.

Their obvious drawback is that they cannot switch off the lights in the small hours. This seems a waste of money, though there are arguments for keeping the lights on anyway (see below). But in both Britain and the US, the power companies sell electricity to local authorities at a bargain price in the early morning. Domestic customers can also have electricity at a fraction of its daytime price. At that time of day, power stations are providing power virtually for nothing – if one ignores the capital cost – so the electricity companies would prefer to get some return on their investment rather than none at all. They say this is more cost effective than shutting down the power stations for the night. The extra cost of keeping a typical residential-area streetlight burning all night, compared with switching it off at about midnight, is no more than roughly £2 (US$3) a year. (Although, of course, this has to be multiplied by a rather large factor to work out the total annual cost of all-night streetlighting.)

Even so, cost is not the only consideration. Most people, if asked, would say they prefer the streetlights to be on all night if it means safer streets. There is little doubt that streetlighting makes the roads safer for drivers and the emergency services, and it could be argued that if one life is saved each year then the expense is well worth it. So by sticking with photoelectric cells, civic authorities are apparently keeping most of their customers happy, which means a great deal to them. Some take a rather mercenary approach to the cost benefit analysis of streetlighting, for example setting the cost to the community of keeping streetlights on against the hospital bills for accidents caused by turning them off. Authorities, particularly in the US, are concerned about their possible financial liability should a citizen be able to prove that an accident or break-in was a result of a streetlight being turned off.

Streetlights and crime

While streetlights on main roads are intended to make driving safer, those in residential areas have an additional purpose – to make the streets feel safer to pedestrians. Local news bulletins and newspapers are dominated by stories of increasing crime, from rapes and muggings to burglaries and car theft. Even in areas where the worst crime for years has been someone riding a bike without a rear light, many people – particularly the elderly – are afraid to go out at night.

Streetlights are seen by the average householder as a defence against this increasing lawlessness, and it would take a powerful argument indeed to persuade them otherwise. The facts are that most break-ins happen during the day, while most crimes against the person are committed in well-lit areas where victims are easy to find. In the UK, a study by the Home Office confirmed that intending criminals pay little attention to the presence of streetlights, but also found that people feel safer with streetlights on though in fact they have little effect in reducing the crime rate. It is no use astronomers pointing out that lack of streetlights does not cause crime – in any debate, they will lose hands down to the majority view. It also matters little that most residential streets these days are deserted after dark.

Positive action

All this makes it sound as if the best thing the city astronomer can do is to head for the country or take up another hobby. But there are ways of lighting our streets and buildings to make them safer without adding to light pollution.

As the public likes to see streetlighting, there is no point in you demanding that streetlights be switched off or taken down – you will be dismissed as a crank. It is much more effective to ask that lighting should be used efficiently, so that not only are the streets kept safe, but the cost is kept to a minimum. There is some public sympathy for the notion of dark skies, so, all things being equal, many authorities will be quite happy to install astronomer-friendly lighting. Although full cutoff lighting is not usually the first choice of lighting engineers, the argument that astronomers have been putting forward for years – that it actually saves money by directing more light onto the ground – is gradually coming to be accepted. The conversion is happening slowly (many lighting engineers have yet to be convinced that full cutoff lighting can actually save money), but it is happening.

National publicity is raising the public awareness of light pollution, but only local action can get things done at neighbourhood level. As this chapter has shown, however, it is important to have facts and figures available which prove your point, otherwise lighting engineers will simply dismiss your demands. Get them on your side, perhaps by appealing to their professional skills to help you in your need for darker skies. Join the International Dark-Sky Association or the Campaign for Dark Skies, and ask for their help first.

Every council and local government authority has an official, often a professional lighting engineer, who is responsible for the public lighting in the area and who sees all planning applications for new schemes. This is the person to whom you should put your point of view. Ask for full cutoff lighting to be installed wherever it is practical, and ask also that consideration be given to switching off non-essential lights after midnight. Your appeal may not have an immediate effect, but it will have raised the issue. When next the engineer reads an article about light pollution in a trade journal, it will ring a bell. Put it this way: if there were no demand for darker skies, no one would do anything about it.

Your local lighting engineer, however, does not have control over all the lighting in the area. Lights on private property are often installed with little professional skill. Once they are up, there is a strong economic argument against removing them. So, if possible, keep an eye on all planning applications which may involve lighting – in the UK they are usually published in local papers – and make your views known in advance. If a new supermarket or shopping area is being planned, ask for details of the proposed lighting.

As proof that things can be done, consider this tale from space science lecturer Richard Taylor. He often uses the observatory of the Croydon Astronomical Society, which is sited within the suburban sprawl south of London, and has helped provide a considerable amount of extra equipment. When it was learned that a new road was planned to run nearby, it seemed that all his and the society's work might be wasted. Nevertheless, a case was submitted which emphasized the educational use of the observatory, requesting that dark-sky standards be applied to the lighting – which in the UK means full cutoff high-pressure sodium. To their delight, their case was accepted without question. They were told that this was because they had made their views known in good time, before detailed plans were made, so they could be taken into account.

In the case of security lighting, you could ask for it to be controlled by infrared beams so that it comes on only when there is an intruder, rather than being left on permanently. This is a contentious area. People who feel they need security lighting do not want to be told by others how to control it. They could argue that the very presence of bright lights puts people off (though there is some evidence to the contrary). Only if the lights are tungsten (white) can they be switched on instantly – sodium or mercury lamps take some time to reach full brilliance.

Miss no opportunity to put your views across. Write to your local papers pointing out that it is up to all of us to control lighting so that the skies remain dark while the light that is needed goes where it should. Write to your MP or Congressional representative about it. Write to the managers of local stores with bad lighting, asking that in the cause of good customer relations and a green image they could gain useful publicity by improving the quality of their fittings.

There have been several cases of bad publicity in the UK, notably when the supermarket chain Tesco opened a new store in 1992 in the former Hoover Building in West London. They floodlit the exterior with brilliant green lighting – actually mercury lights with the blue light filtered out, making it doubly inefficient. They wanted to make an impact but they only succeeded in getting the building dubbed "The Slime Palace" by locals who found it impossible to sleep at night on account of its brilliance. There were numerous traffic accidents on the main road outside. Even non-astronomers complained about the unearthly green glare that destroyed the view of the sky at night. The company soon got the message, and agreed to dim the lighting and to control the amount of time for which it was switched on.

As well as corporate bad lighting, your immediate neighbours may have interior or exterior lights that ruin your observations. Sometimes the perpetrators of bad lighting are totally unaware of the nuisance they are causing. A tactful approach pointing out the problem can sometimes work wonders. It might be a good idea to invite them round to look through your telescope, if you have one, to see a few of the wonders of the skies and to discuss ways in which controlling their lighting could help you see more.

After you have done all this, your skies will probably still be bright. You will have to come to terms with the sad truth that light pollution will be around for a long time to come. So the next thing the city astronomer must do is to choose the right weapons and targets to make the most of what is still available.

CHAPTER 4

Choose your targets

City astronomers must be more selective than their country cousins in what they observe. Generally speaking, the brighter objects can be seen just as well from the city as from the country; it is the fainter objects that cause the problems. The determined city astronomer nevertheless has much to choose from, including many of the sky's showpieces. In this chapter I describe what you can observe from under light-polluted skies, in each case giving a brief guide to getting the most out of your observations – if you want to. There is no reason to feel that you have to look at everything there is to see. To be honest, the vast majority of amateur astronomers rarely observe seriously, but simply get a kick out of viewing their favourite objects. I shall run through the celestial targets in a somewhat arbitrary order of brightness and interest, starting with the Sun and the Solar System.

The Sun

As an object for study, the Sun has a lot going for it. You lose no sleep observing it, you can usually keep warm, it is there in the sky more often than many other astronomical bodies, you do not need a large telescope, and its features change daily, if not hourly.

Being in an urban area is no real drawback to the solar observer. You are as likely to have a building in the way of your target as the country observer is to have a tree, and the seeing can be bad in either location. While it is true that the best seeing demands mountaintop or lakeside sites, the average city location is probably not much worse than the average country one. The best seeing is usually in the morning. Bad industrial pollution reduces the contrast of the Sun's image, but not normally by so much that you cannot observe. All in all, however, the Sun is an ideal target for the city astronomer to study.

The Sun is so bright that a large telescope is a liability rather than an asset. A telescope as small as 50 mm (2 inches) in aperture is adequate, and 75 mm (3 inches) is often said to be ideal. The problem is cutting the light down rather than capturing enough of it, which is the difficulty with most other celestial objects. The Sun is unique in presenting dangers to the observer. Looking at it directly through a telescope or binoculars is like using a burning glass on your eye, with temporary or permanent blindness the result – as Galileo discovered.

Even looking at the Sun with the naked eye is dangerous, and our brains usually prevent us from doing so when it is too bright. Only when it is not painful to look at – such as when it is very low in the sky with its light partially dimmed by haze – is it safe to observe without a filter. There

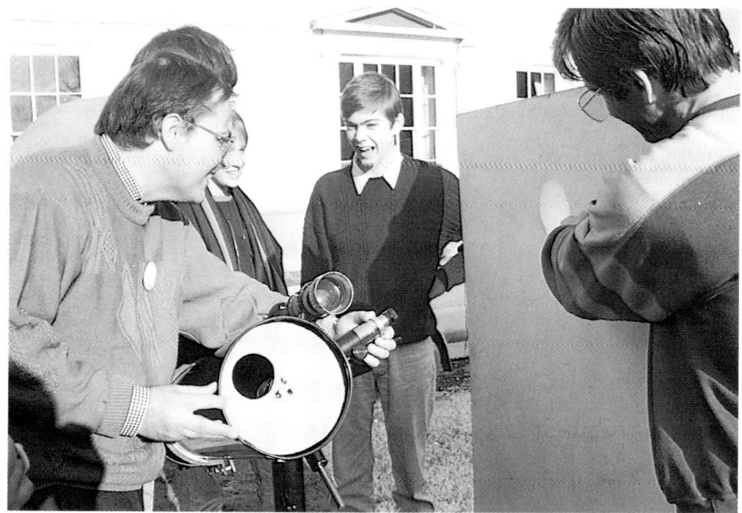

Figure 4.1 *When projecting the Sun, stop the aperture down to around 75 mm (3 inches) to avoid excess heating of the eyepiece. With a Newtonian reflector, as here, cut the hole in the stop so that the Sun's light misses the secondary mirror support. But spot the mistake here – the image from the uncapped finder telescope is in danger of burning someone's sleeve.*

are records of the visibility of sunspots in such conditions, notably from the Orient, from long before Galileo first drew the attention of Renaissance Europe to them in 1610. Today you may be tempted to use a dark filter, but you should steer clear of anything which is not specifically designed for solar work. There is no guarantee that a filter that cuts down visible light will also cut down the infrared, which will burn your retina in the absence of the involuntary reaction against a bright light that would otherwise close the eye.

I must now deliver the obligatory warning against using the solar filters provided with many small telescopes. These are made of dark glass and generally screw onto the eyepiece. There is a grave danger that they will shatter unexpectedly, especially if they are used with the full aperture of the telescope. Usually supplied too is a cap that fits over the top of the tube and cuts down the aperture to about 60 mm (2½ inches) for observing through the solar filter. Experts advise that you should never observe with an eyepiece filter. It is positioned at the greatest concentration of the Sun's rays, so it inevitably heats up. Even a hairline crack, which you may not notice, can let through dangerous amounts of heat. This applies to naked-eye use as well as telescopic work – although a filter is unlikely to crack when simply held up to the Sun, it may be cracked already.

Just pointing your telescope at the Sun can cause problems. Always keep the lens cap on the finder telescope, if you have one, to prevent accidents. Bear in mind that the beam from the eyepiece of either finder or main telescope can burn. The best way to align your telescope on the Sun is to watch the shadow it casts and move it until the shadow is as small as possible.

The safest and cheapest way to observe the Sun is to project its image onto a screen, as in Figure 4.1. You can do this with any kind of telescope. Some instruments come equipped with solar projection screens. Regular observers use boxes which keep stray light off the screen, giving much better contrast. To get the most spectacular view, project the Sun's image into a blacked-out garden shed. Rather than pointing the telescope directly at the Sun, use a surface-silvered flat mirror placed outside the shed to reflect the Sun's light into the telescope, which is fixed horizontally. An image about half a metre across can be projected in this way.

This method does have its practical difficulties. First, the mirror needs to be much flatter than ordinary glass, and ideally made of low-expansion material, which usually means that it has to be made to astronomical specifications. To track the Sun across the sky, you will need an arrangement for turning the mirror at the rate of one revolution in 48 hours – half the rate provided by standard astronomical drives. But if you can overcome these problems, you will be rewarded with fine views of our nearest star. Sunspots stand out with good contrast, and the bright areas known as faculae also show up well (see Figure 4.2).

Solar projection allows you to make drawings of the Sun's features, which is the usual means of recording their positions. Not only can you detect the Sun's rotation, but you can also see how the spots change from day to day. For drawing the Sun, standard blanks with a diameter of 152 mm (6 inches) are available. It is possible to mark the positions of features on a blank disk by simply projecting the Sun's image onto it and marking them in, but a better method is to project the image onto a faint pre-prepared grid. You then transfer the locations of the features onto a

Figure 4.2 *The Sun photographed in white light, showing several sunspots. This view was obtained by photographing the projected image in a darkened room, using an ordinary camera alongside the telescope.*

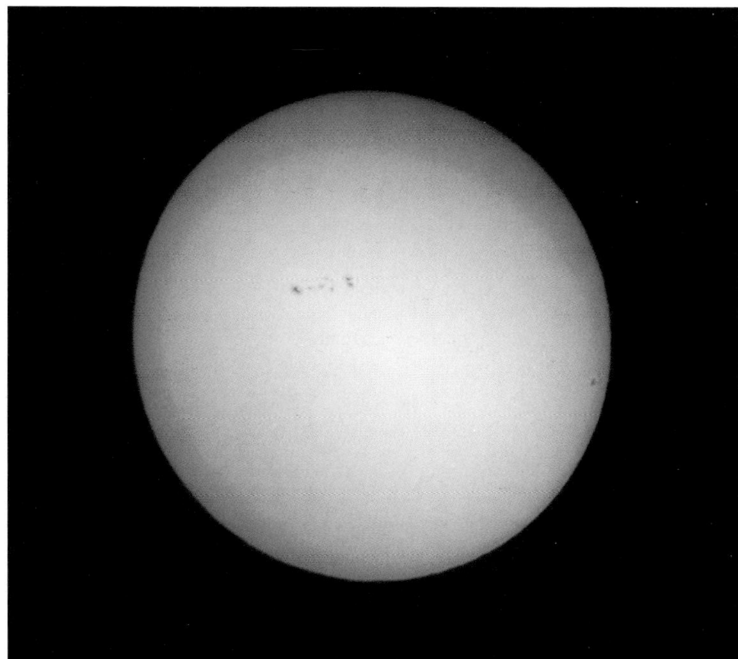

blank disk under which you have put an identical grid ruled with bold lines that show through the paper.

An alternative method of observing the Sun is to place a filter over the objective of a refractor, or over a hole cut in a mask fixed securely at the top end of a reflector. The cheapest of these filters are made from thin aluminized Mylar film, and are definitely not suitable for use at the eyepiece end. The film is delicate, and two thicknesses may be needed. Before using such a filter, check that there are no tiny pinholes in the coating that could let through light. Most important, make absolutely sure that the filter cannot be dislodged under any circumstances. Using such a solar filter, particularly a home-made one, is rather like driving at speed in the fast lane – it is perfectly safe most of the time and you can feel quite secure, but if the unexpected happens the results can be sudden and disastrous. More expensive filters also use reflective coatings, but on optical-quality glass mounted in a cell which can be firmly fixed to the objective. These can also be damaged by rough handling. Always follow the manufacturer's instructions to the letter when using solar filters.

Some observers are always trying to find cheap ways of filtering the Sun's light. Food packaging made from aluminized Mylar should not be used to make a solar filter: it was intended to protect the food, not your eyes. I have seen pieces of coloured plastic offered for sale as solar filters with extravagant claims that could not be backed up. When one of these filters was put in front of a TV remote control, which sends signals at infrared wavelengths, the TV still responded to the commands, showing that the filter transmitted at least some infrared. This is by no means a definitive way of selecting materials opaque to infrared – that can be done only by detailed laboratory testing. It did reveal, however, that there are no established guidelines for safety when observing the Sun. Welder's glass must be regarded with suspicion since it is designed for a different purpose. However, for inspecting the Sun without a telescope to check for naked-eye sunspots or when viewing an eclipse, a Number 14 welder's glass can be recommended.

It is not necessary to draw the Sun each time you observe it. You can if you wish simply count the number of active regions and sunspots. An *active region* is an area in which there is a single large spot, or several spots all within 10° of one another in terms of solar latitude and longitude. To begin with, you need to record the positions of all the spots visible and then work out how many are within 10° of each other. With practice you can do this simply by estimation.

The daily tally of active regions and sunspots may be used to work out the Sunspot Number. Simply multiply the number of active regions by 10, and add the number of sunspots – spots belonging to an active region as well as lone spots. So if there are two active regions and seven individual sunspots in all, the Sunspot Number for the day is 27. Estimates vary from person to person, and with the equipment used (close spots may be difficult to distinguish). You can compare your results with those published in *Sky & Telescope* each month.

The way the Sunspot Number changes over the months and years is of more than academic interest. The number varies, with a very obvious cycle of roughly 11 years. At times there are few, if any, spots visible on the Sun, and those that can be seen are all at high solar latitudes. A few years later the Sun will be covered in spots, mostly closer to the equator. At the peak of the solar cycle the spots are at their most numerous and other solar phenomena also reach a maximum. At these times, frequent bursts of radiation from the Sun can have a significant effect on Earth. Aurorae (the northern and southern lights) are much more common, and occasionally long-distance radio communications and even power supplies are disrupted. There may be extra drag on satellites in low orbits, causing them to spiral down to a premature end. Astronauts in space may be subjected to additional radiation hazards.

The Sun at other wavelengths

Straightforward white-light observation of the Sun's disk, as described so far, is the bread and butter of the solar observer. But enthusiasts have long made use of the ample supplies of radiation from the Sun across the whole electromagnetic spectrum to single out other wavelengths for study.

The Sun's visible spectrum, like that of most other stars, consists of a bright rainbow of colours crossed by dark lines produced by atoms and molecules in the Sun's atmosphere. These lines are named *Fraunhofer lines* after the German physicist Josef von Fraunhofer, who was the first to map and study them at the beginning of the nineteenth century. Hydrogen is by far the most abundant element in the Sun, and the three strong hydrogen lines in the visible part of the spectrum are favourite targets for solar observers. The lines are the first in a series of lines called the Balmer series, and are assigned letters of the Greek alphabet. *Hydrogen-alpha* (usually called H-alpha) has a wavelength of 656.3 nm, corresponding to a deep red colour near the limit of the human eye's red sensitivity. It can therefore be hard to spot in the Sun's spectrum, even though it is a very strong line.

Although the Sun's atmosphere absorbs light at this wavelength, there is still plenty that gets through. The giant protrusions from the Sun's atmosphere known as prominences actually shine with this colour. They come into view during total eclipses of the Sun, looking like pink flames around the limb. The pink colour is a combination of the deep red hydrogen-alpha, the less strong green hydrogen-beta, and the weaker blue hydrogen-gamma line.

At one time, if you wanted to see prominences without waiting for a solar eclipse you had to make a spectroheliosope – an instrument for isolating a particular line from the Sun's spectrum and scanning its disk in the light from this one line. Such devices are versatile, since they let you observe using any wavelength you choose, but they require a fair amount of effort and space. These days, however, you can buy filters which transmit hydrogen-alpha only. The cost is not trivial: for the filter alone, expect to pay about the same as for a basic 150-mm (6-inch) reflector. There are even more expensive versions that transmit a narrower wavelength band. Do not mistake these *narrowband filters* for those intended for deep-sky photography – also referred to as H-alpha filters – which simply cut out all radiation other than the deep red. Note too that the exact wavelength transmitted by some filters depends on their temperature, and may also vary with the filter's age. Beware of second-hand filters that may need to be heated to extreme temperatures before they transmit H-alpha!

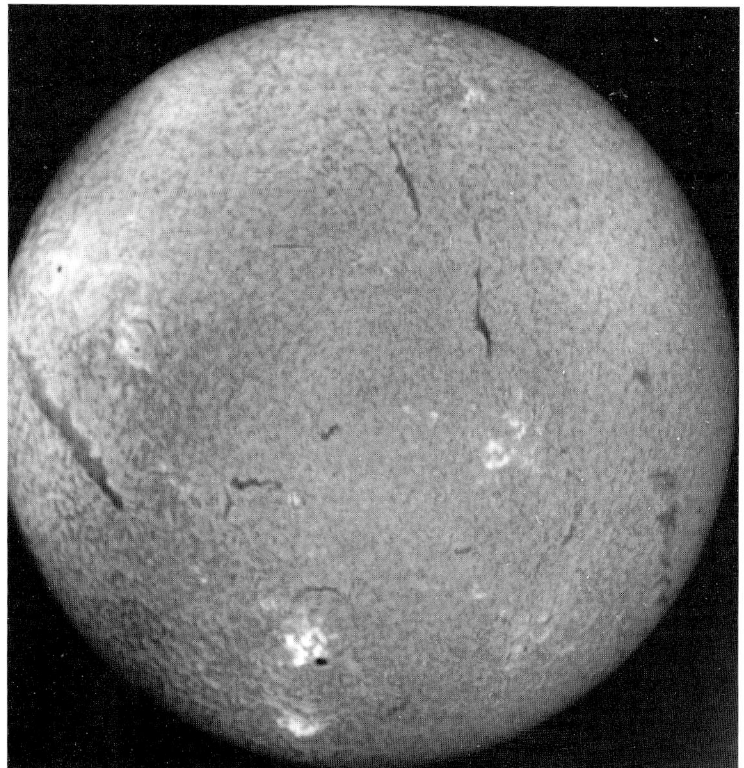

Figure 4.3 *A view of the Sun through a hydrogen-alpha filter with a bandwidth of 0.06 nm, taken by Eric Strach of Liverpool, UK.*

With an H-alpha filter you can see prominences around the limb of the Sun, if any are present. Most of the time you will see only a few rather insignificant spikes sticking up, or maybe nothing at all. They may change shape in a matter of minutes, and may detach themselves altogether from the limb. Do not expect to see anything like those movie sequences of giant flames leaping from the Sun. They were taken at times of extreme solar activity, and have been greatly speeded up. The most spectacular prominences are usually associated with large sunspots.

The disk of the Sun as seen in H-alpha has a much more mottled appearance than in white light (see Figure 4.3). The sunspots are not so easy to see, but their locations are usually obvious from the more disturbed nature of the area. As well as H-alpha filters you can get calcium filters, which isolate the blue calcium line.

The Sun can also be observed at radio wavelengths, a topic covered in Chapter 8.

The Moon

People expect a lot from their first view through an astronomical telescope. They want to see the dramatic swirls of gas clouds in space, the vivid festoons of colour on Jupiter, the fantastic landscapes of alien moons. In most cases, the reality is disappointing by comparison – except for just two objects. One is the Moon, especially at first quarter. (The other is Saturn – see page 54.)

No matter what instrument you use, from binoculars to a high-power telescope, the Moon can be a spectacular sight. It is a challenge to try to observe the Moon when it is a thin crescent, which demands a good horizon and clear skies at just the right time. Just after sunset light pollution does not interfere, though urban pollution does. An observer in a high building stands a better chance than one at ground level, so a suitably located city observer could have more success than one in the country. Binoculars are helpful in picking out a very young Moon. The real test is to see a Moon which is less than 24 hours old.

Telescopically, the Moon provides hours of enjoyment. I like to gaze along the limb, where I can see the true slopes of the mountains, as if I were on the surface itself. The harsh lighting and long shadows of the *terminator* – the shadow line – give the impression that the Moon's landscape is more rugged than it really is (see Figure 4.4). At the limb, you can see that the slopes are actually gentle and rolling, rather than jagged.

The Moon is undeniably a crowd-puller, but what can you actually do with it? A generation or so ago, there were still parts of its visible face waiting to be mapped by non-professionals. Amateurs were still making discoveries, and hotly debating the origin of the Moon's features. Now virtually the entire Moon has been mapped, men have walked its grey rolling plains, and scientists have analysed its rocks in Earth laboratories. Is there anything left for the amateur to contribute?

Surface changes?

There is one very good reason for watching the Moon. Occasionally – very occasionally – odd things seem to happen, for there have been reports of unusual glows or obscurations. These are grouped under the general heading *transient lunar phenomena*, or sometimes lunar transient phenomena (TLPs or LTPs). Like reports of UFOs, the Yeti, or Bigfoot, they can seem reliable or even convincing, but are tantalizingly hard to explain. Are they escapes of gas from the interior, or the result of meteorite strikes, or simply illusions? We still await more undeniable events, seen independently by several observers.

There are certain craters, Aristarchus, Grimaldi, and Plato among them, whose names keep cropping up in TLP reports. The Aristarchus area in particular, including the nearby Herodotus and the "Cobra Head" of Schröter's Valley, has given rise to a third of all TLP sightings. This is a fascinating area of the Moon – Aristarchus is the brightest lunar crater, and one of the most recently formed. In 1963 reddish glows were reliably reported near Aristarchus, and a brightening was photographed during a solar flare. Spectrograms have been recorded of other glows, leaving little doubt as to their reality.

Should you ever see an unusual phenomenon, you should alert other observers if possible, but avoid giving them clues as to what is happening, or exactly where, to prevent bias in their reports. Observations of colour are best substantiated using a reflector or a top-quality, perfectly colour-corrected refractor so as to avoid any suspicion that the colour is an artifact of the instrument. A filter of contrasting colour can help to enhance the contrast of any colours seen – a blue filter will make a red patch appear unusually dark, for example. A pair of red and blue filters, used in succession, can make a coloured area appear to blink in brightness. The British

Figure 4.4 *With almost any form of optical aid the Moon is a spectacular sight, particularly when close to a quarter phase when the terminator region is thrown into sharp relief by the low angle of illumination from the Sun.*

Astronomical Association's Lunar Section and the Association of Lunar and Planetary Observers collate observations of TLPs.

Although it is worth keeping an open mind about such phenomena, it has to be admitted that very few people have ever seen a TLP, and that to all intents and purposes the Moon's face is unchanging. But there is another good reason for observing it: it is ideal for honing your observing skills. In these days of high-tech photography and CCDs it may seem quaint to go out and draw part of the Moon, but it is an excellent way of training your eye.

Set yourself a task – mapping a certain area, for instance. Make a detailed drawing every so often, and you will find that your perception of the different features changes. What seemed at first to be a hillock turns out later, when the lighting is different, to be a craterlet. The Moon rarely presents exactly the same face to us each month, but wobbles so that we see slightly more of one side one month, then slightly more of the other side the next. This effect is called libration. The Sun's angle above your chosen area will also vary, from, say, one first quarter to the next, so you can get a huge range of aspects of just one small part of the Moon over a period of time.

Occultations

There is another facet of lunar work which can actually help contribute to knowledge, and that is the observation of lunar occultations. Basically, this means watching as the Moon passes in front of a star, and timing the exact instant at which the star disappears or reappears. This may seem trivial, but it is both more difficult than it sounds and a rarer event than you might imagine. There are surprisingly few stars in the path of the Moon that are bright enough to be seen against its glare – usually about a dozen each month are observable from any one location. Most of them are quite faint and need good equipment (though not necessarily large apertures) and very clear skies to be seen at all.

Your occultation timings must be highly accurate, and your reaction time will be a major source of error. Your telescope should have a long *f*-ratio and give good contrast, so a 75-mm (3-inch) refractor may be better than a larger Schmidt–Cassegrain (see Chapter 5). A digital watch with a stopwatch facility is quite adequate for timings, but you must check it before and after the event against an official time signal, such as a radio time signal or telephone time service. Casual timings are of no use.

It is hard to see any but the brightest stars at the bright limb of the Moon with certainty. Most observations are therefore of occultation events taking place at the dark limb of the Moon – that is, disappearances well before full Moon and reappearances well after full Moon. The good news is that you can usually observe lunar occultations just as well from the city as from the country. Wherever you live, the chief requirement is a good clear sky. The light from the Moon itself is the major contributor to the sky brightness.

The Moon's track as seen in the sky depends on whereabouts on the Earth's surface it is seen from, so predictions of occultations are published for each country by the International Occultation Timing Association. For Britain, Australia, and New Zealand a brief list appears in the annual *Handbook* of the British Astronomical Association (BAA), while the annual *Observer's Handbook* of the Royal Astronomical Society of Canada lists events for the US and Canada. These organizations also collect observations and relay them to the International Lunar Occultation Centre in Japan. The results are used to refine our knowledge of the orbits of both the Moon and Earth, and are of genuine scientific value.

Occasionally a star may not be completely hidden during an occultation, but the Moon's upper or lower limb just skims past it. This is known as a *grazing occultation*, and is particularly useful and interesting to observe. The star may appear to wink off and on several times as it is glimpsed through valleys on the limb. You need to be in just the right place to see this happen, and it usually means travelling some distance from your home. The Moon's path in space is not known precisely, so there is usually some uncertainty about the track on the Earth's surface along which the Moon will just be seen to graze the star. Ideally there should be several observers spread out a mile or so either side of the predicted graze line, at right angles to it. Some observers will probably see no occultation at all, while others will see a brief occultation. Some will be lucky enough to see the graze. Portable telescopes are quite adequate.

Mars

Of all the planets, Saturn notwithstanding, Mars is the one that fires the imagination. With its sandy deserts and thin atmosphere, it is the only place beyond the Earth where humans might feel even remotely at home. Even through the telescope Mars is usually disappointingly far away – it is a small world, and comes within good observation distance for only a few months every two years or so at the time of opposition (that is, when its orbit brings it closest to Earth). Mars has a very elliptical orbit, and its distance from the Earth at opposition varies by a factor of 1.8. The net result is that the planet is best placed for amateur astronomers only at intervals of 15 and 17 years.

At a favourable opposition even a small telescope will show some detail on Mars, as Figure 4.5 shows, though you really need a telescope of at least 150 mm (6 inches) aperture to see anything worthwhile. Your first glimpse could be disappointing. If the seeing is not very good, you may well see two or three Marses, all shimmering and overlapping. Once things have settled down you will at first see a small orange disk and little else, but persevere and you will notice a distinctly darker shading, and maybe a hint of a white polar cap. You may find that from time to time the seeing improves sufficiently for you to see quite a bit of detail. If you have been observing for some time, perhaps making a drawing, you begin to realize that the markings are no longer in the same places as when you first observed them. Mars turns on its axis at much the same rate as Earth – in 24 hours 37 minutes, in fact – so bringing new features into view. If you observe at about the same time the next night, you will see the same part of the planet.

There is quite a range of features to be seen on Mars: dark and bright areas, polar caps (of which only one is usually visible at any time), and clouds of varying appearance. Strong winds blow through the thin atmosphere, whipping up the dust that covers much of the surface and spreading it around. As the underlying rocks are covered and uncovered, the appearance of the markings we see from Earth changes from opposition to opposition, and sometimes from week to week. Such dust storms can

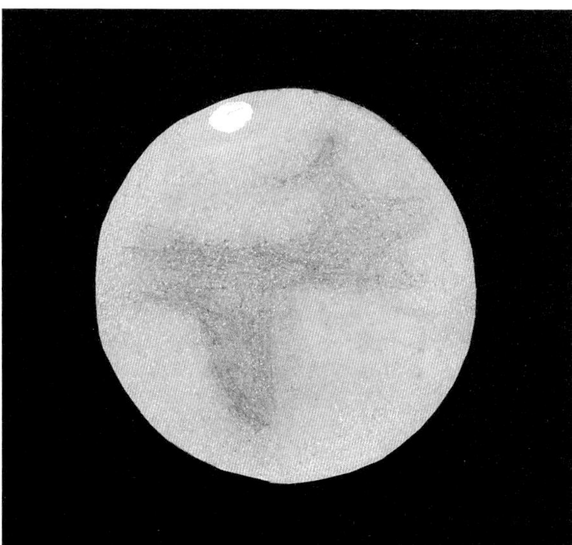

Figure 4.5 *Mars, sketched by the author in September 1988; the telescope used was a simple 90-mm (3½-inch) reflector with a plastic drainpipe for a tube. The drawing shows some detail, despite the very basic telescope and limited artistic ability.*

blow up with no warning. I remember taking a look at Mars from an observatory on top of a college building in the middle of Manchester in 1971. I was rather disappointed with my view through the 150-mm (6-inch) refractor, as I could see no detail at all. Shortly afterwards I heard that a major dust storm had just begun, coincidentally just as the Mariner 9 space probe entered orbit to begin photographing the planet. Despite my initial impression, my city location had been no hindrance to my observation of the planet.

It is the shifting sands of Mars and the planet's constantly varying appearance that make it so fascinating to watch. As it turns at roughly the same rate as Earth, observations you make could complement those made by Mars observers in other parts of the world, who will be seeing a different part of the planet at a different time. You could be the first to catch the start of a new dust storm. The way the planet changes its appearance is of interest to planetologists who are studying the climate of Mars. After all, one day humans will certainly set foot on Mars, and amateur observations made now will add to the knowledge of the sort of weather they can expect.

Dust storms were once thought to occur mostly when Mars is at its closest oppositions, when the planet is also closest to and heated most strongly by the Sun. But analysis of observations made since 1873 has shown that they can occur at almost any time, with only brief dust-free spells. One would imagine that with the prospect of astronauts visiting Mars in the not too distant future, NASA would be funding a continuous watch on the planet by professional observers, but this task is largely being left to amateurs. The Association of Lunar and Planetary Observers is conducting a long-term Mars patrol and welcomes detailed observations of the planet made at all times, not just at opposition.

Mars is a notoriously difficult planet to observe away from the times of opposition, because its diameter is very small: only about 4 arc seconds when on the far side of its orbit from Earth. Nevertheless, amateur observers are now able to make worthwhile observations by using CCD electronic imaging systems. Once the image is on disk, image processing software can be used to bring out previously unseen detail. Don Parker, of Coral Gables, Florida, for example, produces superb planetary images using a 410-mm (16-inch) *f*/6 reflector equipped with a CCD.

Jupiter

Whereas Mars is small and infrequently well seen, Jupiter is a regular sight in our skies and big enough to show detail in the smallest of telescopes. The planet's brightness means that Jupiter observers can study their target in light-polluted and murky skies that would make other observers pack up. Like Mars, Jupiter presents us with ever-varying features, and its appearance can change significantly from year to year. But while Mars is a rocky planet, Jupiter is a gas giant with no observable solid surface. What we see is the changing patterns in its thick upper atmosphere. There are dark *belts*, fascinating to study as they can suddenly fade, causing a radical change in the appearance of the planet. Separating the belts are the lighter *zones*. The Jupiter observer must get to grips with the terminology of these belts and zones, which features such tortuous descriptions as North North Temperate Belt (South).

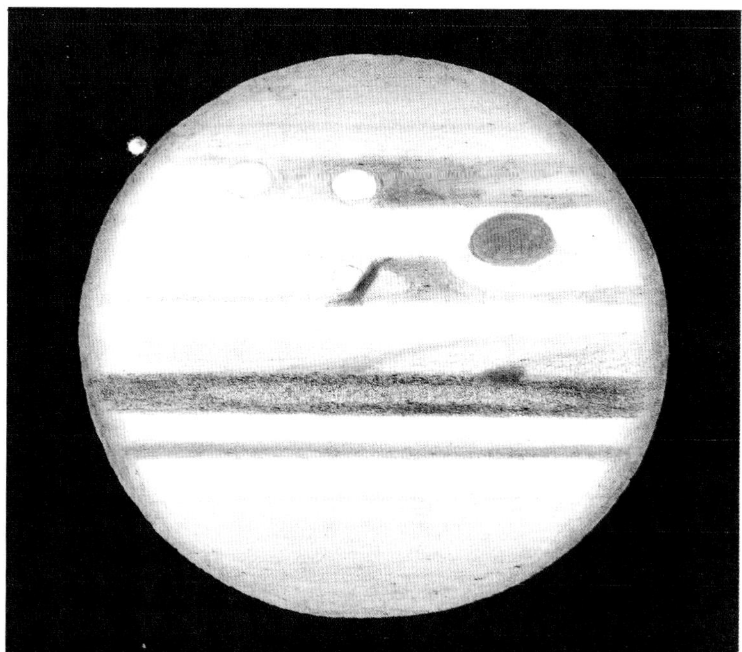

Figure 4.6 *A drawing of Jupiter by John Rogers of Cambridge, UK, prepared on a standard blank supplied by the British Astronomical Association. It is not necessary to use colour when drawing a planet, though in the case of Jupiter there are various hues to record, of the Great Red Spot in particular.*

Study Jupiter and you will see that the edges of the belts are not nearly as straight as they appear at first glance. Swirls of gas lap into the neighbouring zones, and white spots and ovals come and go. There is a special nomenclature for these features: Jupiter observers speak of such things as "projections," "festoons," and "barges." Colour is immediately apparent – mostly shades of cream and yellow, but also browns, reds, and blues. Most striking is the famous Great Red Spot, which has been a permanent feature ever since Robert Hooke first reported it in 1664. It changes its colour over the years, sometimes becoming so pale as to be virtually white. More often it is a salmon pink, and sometimes it takes on a true red hue so as to justify its name.

Jupiter rotates very rapidly – in less than 10 hours – so not only do the features move round the disk quite noticeably as you observe, but the planet is clearly flattened. As Jupiter is not a solid body, its rotation period varies with latitude, with about five minutes a day difference between low and high latitudes. The rapid rotation means that, as with Mars, the planet presents quite a different face from that seen by an observer in another country a few hours later or earlier. Space probes and the Hubble Space Telescope may have shown us much more detail than is visible from the Earth's surface, but they cannot always provide the round-the-clock monitoring that amateur observations can.

The ever-changing disk of Jupiter presents observers with a unique challenge. Minor features come and go, and also drift in longitude, borne by atmospheric currents. Even the belts and zones have occasionally undergone changes. The problem is to record as much as possible sufficiently quickly before the planet's rotation takes some features away and brings others into view. A straightforward drawing of the planet is the standard method of recording its appearance. The drawing need not be artistically pleasing as long as the features are accurately plotted. A careful record of the view through a 150- or 200-mm (6- or 8-inch) telescope can show a wealth of detail. Experienced observers can produce drawings very quickly, and with time the ability to reproduce fine shadings improves. Because the planet's disk is so flattened, Jupiter observers use prepared blanks of the right shape. The drawing shown in Figure 4.6 was prepared on such a blank.

The usual way of measuring accurate positions for features is to estimate the time when they cross Jupiter's centreline – the *central meridian* – which is fairly easy to visualize because of the planet's flattening. Timings to the nearest minute correspond to an accuracy of 0.6° of longitude on Jupiter. Such measures can then be used to produce charts showing the changes in longitude of features that result from the different rotation rates and motion in the atmosphere. For example, it is quite common for spots or ovals to change their rotation speed suddenly.

If drawing the complex appearance of Jupiter does not appeal to you, photographs or CCD images are a good alternative – though photographing any of the planets is a much harder task than it might appear. Further details are given in Chapter 6.

The smallest telescope or even binoculars will reveal the four largest moons of Jupiter, known as the Galilean satellites as they were first reported by Galileo. The constant shuttling of these moons – among the largest in the Solar System – is a perennial fascination. It is hard to see them as disks unless you have a large telescope and good seeing, but the shadows they cast on Jupiter as they cross in front of it are easily visible. The moons can also be seen going into or coming out of eclipse. Precise timing of these events can be of scientific use, though this is best done using a photometer rather than with the naked eye.

Saturn

Were it not for its glorious ring system, Saturn would be a very neglected planet. It is smaller and more distant than Jupiter, and its disk shows very much less detail. The most exciting thing that ever happens is the outbreak of a white spot every 30 years or so. But the rings turn a distinctly saturnine planet into one of the true spectacles of the sky. In London's West End there is often a street astronomer who shows passers-by some celestial lollipop on clear nights. Those theatregoers who pause to gaze though his telescope when Saturn is glimmering over the rooftops are in for a treat.

The dedicated Saturn observer, however, must search for more elusive phenomena. The planet's features are more subtle than those of Jupiter, but there is still much to be seen and done, though as with the other planets, the larger your telescope, the better. Like Jupiter, Saturn has dark belts, and these vary in position and brightness. Measuring their positions and making estimates of their brightness keeps Saturn observers busy. Some small white spots are occasionally seen, and observers also make notes of the colours of the belts and zones. Colour filters can help to emphasize any slight colour differences.

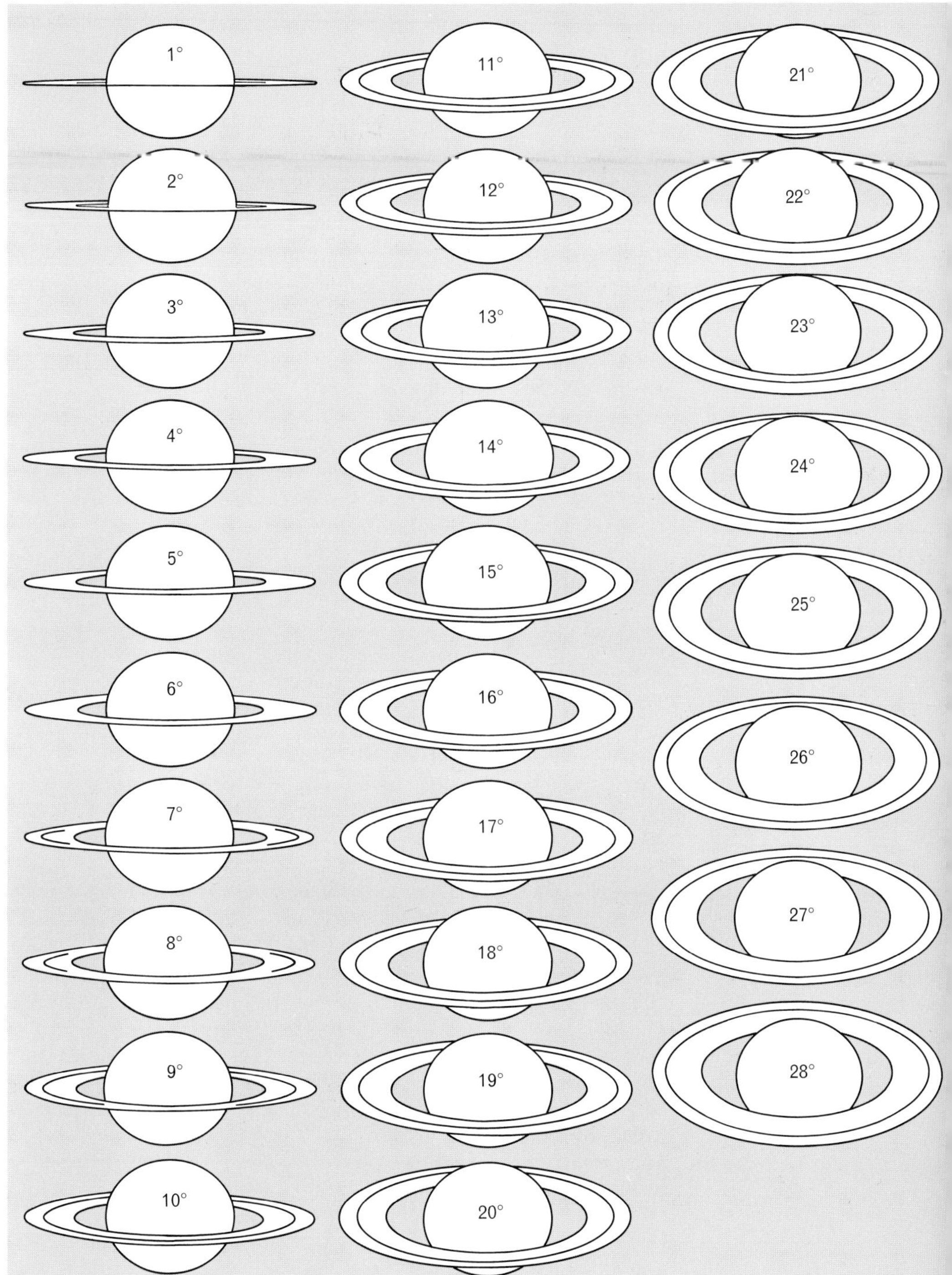

Figure 4.7 *Saturn changes its orientation throughout its 29½-year orbit, as shown in this sequence taken from the BAA Saturn Section's set of observing blanks. Depictions of the tilt of Saturn for each month are published in* Sky & Telescope's *"Stars and Planets" section.*

Figure 4.8 *Saturn, drawn by Matthew Boulton of Redditch, UK, in August 1991. He used a 157-mm (6¼-inch) reflector.*

As with Jupiter, drawings of Saturn are best made on a prepared blank. The problem here is that because of Saturn's high inclination, the angle at which the rings are tilted towards us changes continuously in the course of the planet's 29½-year orbit, so a single blank will not suffice (see Figure 4.7). Figure 4.8 shows a typical observer's sketch of the planet. Every 15 years or so the rings are presented to us edge-on, as in 1995, and it is at these times that it becomes obvious how thin they really are, since they can virtually disappear. At around these times too we can catch a rare glimpse of the unilluminated side of the rings if the Sun is on the opposite side of the plane of the rings from the Earth.

Any telescope larger than about 60 mm (2.4 inches) aperture will show the division between the outer, narrower, and fainter Ring A and the inner, broader, and brighter Ring B. It is known as Cassini's Division after Giovanni Cassini, who discovered it in 1675. When the seeing is really good you might be able to pick out Encke's Division, within Ring A, first seen in 1837 by Johann Encke. Occasionally, observers report suggestions of additional divisions within the rings or faint rings inside or outside the main ones. The shadows cast by the rings on the globe and by the globe on the rings are also interesting to watch, and sometimes irregularities can be seen. To see details such as these you will have to study the planet carefully – rarely does Saturn reveal its secrets to the casual spectator.

An urban location is no handicap to the Saturn observer. The former Director of the BAA's Saturn Section, Alan Heath, lives in a suburb sandwiched between the cities of Nottingham and Derby where the naked-eye limiting magnitude (that is, the faintest star visible) is rarely better than 4. He finds, paradoxically, that very subtle detail, such as Encke's Division, can actually be easier to see when the sky is bright, in twilight. This may be because there is then less glare from the planet's disk. Alan's location is made even worse by a large power station a few miles to the south. When the planet is low in that direction, the seeing can be very bad. Alan finds that his best seeing is in what he calls "pre-fog" conditions, about half an hour before a mist descends and reduces the transparency.

The other planets

Mars, Jupiter, and Saturn are undoubtedly the main planetary attractions for amateur observers, but the inner planets and those farthest from the Sun are not without interest. For all but Venus, though, you need quite a big telescope to do anything more than track their movements.

Mercury and Venus

Mercury and Venus are both closer to the Sun than the Earth is, so are never far from the Sun in the sky. Elusive Mercury is never more than 28° from the Sun, and is therefore usually seen low down in a light dawn or dusk sky. Most amateur astronomers regard it as a curiosity rather than an object for serious study. To be honest, it serves mostly as a test of your observing skill, for you will do well to make out any markings on it at all. For the city observer, Mercury may be a particular challenge if there are local obstructions to the east or west. The best conditions for observing it through the telescope are while the sky is still light, but take great care when searching for it that you do not encounter the Sun by accident.

Venus, however, is much more easily seen. As the evening star it is one of the most obvious of planets, hanging like a searchlight in the sky. This is not the best time to observe Venus, for it is simply too dazzling, and usually too low in the sky for good seeing. Like Mercury, it is best observed while the sky is still light.

But unlike Mercury, there are things about Venus that make it worth observing. Although its surface is hidden beneath a dense cloud cover, there are faint shadings which require clear skies and good, clean optics if they are to be glimpsed. Particularly interesting is the time of *dichotomy* – that is, when the planet is exactly at half phase, with the terminator running straight up and down. Whatever sort of telescope they are using, observers usually find that at the time when the planet should in theory be at half phase, slightly less than half of its disk is illuminated. Another mystery is the *ashen light* – a faint glow on the unlit portion of the crescent Venus, like earthshine on the crescent Moon. Earthshine is caused by sunlight reflected from the Earth and onto the Moon, but there is nothing that can direct sunlight onto the dark side of Venus.

The cause of both the anomalous time of dichotomy and the ashen light probably lies in Venus's dense atmosphere. When Venus passed in front of a bright star in 1981, I and a few others who were observing the event from the centre of the occultation line saw that a faint image of the star clung to the limb of the planet for a few minutes after the time when it should have passed out of sight. The starlight was presumably being refracted around the limb. If this can happen, then it is certainly possible for sunlight to be refracted around the planet to the dark side, which would offer an explanation of the ashen light. Another theory is that the ashen light is caused by the release of energy stored in gas molecules during the daytime.

Since Venus observers generally find that they get the best seeing when the planet is high in the sky during daylight, even the most light-polluted city environment is no hindrance when you are looking for shadings on the planet's disk. The chief requirement is a clear sky. The ashen light does need a dark background to be seen, so you are less likely than your country cousin to see it.

Uranus, Neptune, and Pluto

Not surprisingly, the far outer planets are harder to find than the others because they are so faint. You need a star map with their positions marked, or setting circles, or a computer-controlled telescope to find Uranus or Neptune. With Pluto you need a fairly large telescope, an accurate position, and observations over several nights to be certain that you have actually seen it.

Uranus and Neptune show tiny disks which are usually featureless. I have seen the suspicion of a marking on Uranus with a 450-mm (18-inch) refractor from the London suburb of Mill Hill, so I know it can be done. It is always worth looking if you have a large enough instrument. As for Pluto, you need the Hubble Space Telescope to see even a disk, so just be happy if you find it! It is around 14th magnitude, so you need at least a 200-mm (8-inch) telescope to have any chance of tracking it down. Plot the positions of stars in its vicinity from night to night, and you should find that Pluto has moved.

There is a trick to use when looking for objects near the limiting magnitude of your instrument. If the seeing is good, you can magnify the image up to the limit that your telescope will bear without making star images any fainter, since they are points of light. The sky background, however, gets darker as the magnification increases since its light is being spread over a larger area. So you can increase the brightness difference between the object and the background by using high powers.

Also, faint objects are best seen if you do not look directly at them. Your eye is less sensitive to light right at the centre of the field of view, although that is where it gives the best resolving power. So use *averted vision* – that is, look a little to the side of where you expect the object to be. To avoid the faint image falling right on your blind spot, keep the object to the side of the field of view nearest your nose, whichever eye you are using.

You may find that making the telescope quiver slightly helps you to be sure that what you can see is a star rather than a defect in your eye. Looking hard can help you to see clearly. It has been claimed that light builds up in your eye over a matter of seconds, just as it does on a photographic film, though there is no physiological evidence for this. I suspect that it is the result of concentration and avoiding the small eye movements we unconsciously make.

Asteroids

These minor planets range in brightness from about magnitude 5.5 for Vesta to as faint as you can go. They look just like stars ("asteroid" means "starlike"), except that they change position slightly from night to night. Interest in asteroids has increased in recent years, among both amateurs and professionals. Some of them probably represent the basic material from which the Solar System was formed.

The amateur astronomer can estimate an asteroid's brightness compared with those of nearby stars, just as if it were a variable star (see page 67). If carried out regularly, such estimates can reveal the rotation period of the body. Some amateurs measure the positions of asteroids very accurately, and may even discover new ones – which gives them the right to suggest a name for the newly found member of the Sun's family. Predictions of positions are published in the BAA *Handbook* each year, and also in the RASC *Observer's Handbook*. Computer programs are available that will calculate asteroid positions from their orbital elements.

Until recently, measuring the positions of asteroids, which comes under the general heading of astrometry, was a specialized task, requiring an accurate measuring machine as well as the means of taking good photographs of the objects. The arrival of CCDs (see Chapter 6) in amateur hands could make this line of work very much more accessible. The combination of the new comparatively large-area CCDs and the *Guide Star Catalog* (see page 141) has made it much easier than before to record accurate positions of asteroids, and accuracies of half an arc second are now possible. The positions of stars given in the GSC are not that accurate, but if several are recorded on the same frame an average can be found.

Not all asteroids have the same composition. They can be classified according to their colour, found by measuring their brightness through different colour filters. This work is now within the scope of amateurs using CCDs or photoelectric photometers (see Chapter 6).

It is the fainter asteroids – below about ninth magnitude – which are of particular interest, but even these are observable in city skies with quite modest telescopes. A recent estimate puts the total number of serious asteroid observers throughout the world at 75. With more than 15,000 asteroids known, for most of which only rough orbits are available, there is more than enough room for enthusiasts who want to make this their field.

Comets

Most amateur astronomers, particularly those who live in cities, would unhesitatingly vote comets the most frustrating of astronomical objects, for several reasons. First, they are potentially the most spectacular of celestial sights – yet in the twentieth century there have been maddeningly few. Second, they can be extremely fickle, arriving on the scene with great promise yet delivering virtually nothing. Third, they have a habit of disappearing in the slightest hint of light pollution. While from the city we strain to see a hazy head, from a dark country site the same comet clearly sports a magnificent long tail.

Living in the city is little hindrance to observing the objects described in this chapter so far. But when it comes to comets, the streetlights and pollution can easily beat us. The problem with comets is that they often have a low surface brightness, even though they may span several degrees of sky. Only the immediate surroundings of the head are reasonably bright, and this is all we see in our light-polluted city sky. Even a comet that reaches its predicted brightness may appear much fainter. Published magnitude predictions refer to the total amount of light from the comet and may well imply naked-eye visibility, such as magnitude 4 or 5. But much of that brightness is spread out over a considerable area. The head of the comet is much fainter, perhaps about magnitude 7 or 8. Comets often appear in twilight, low in the sky, since they brighten up only when they are near the Sun. Under these circumstances, the sky brightness can completely drown them out.

Nevertheless, some comets can be seen in urban skies. These are the ones that are condensed, and have the courtesy to travel high in the sky. Figure 4.9 shows an observer's sketch of such an obliging comet. You will need a good, clear sky, a well-baffled telescope (see page 100), and a site well shielded from any nearby streetlights. To find the comet you will probably have to plot its position on a good star map and "star-hop" to its exact location. A few moderately bright comets appear most years, but rarely do they meet any of these requirements. By "moderately bright"

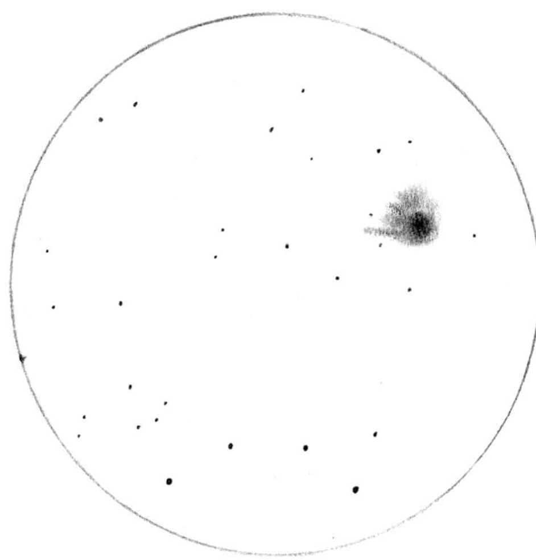

Figure 4.9 *A comet can be observed from light-polluted areas if it is condensed and not low down in the sky. This drawing of Comet Mueller 1993a was made by Matthew Boulton from Redditch, UK, on a night of excellent transparency, with the aid of a 250-mm (10-inch) Newtonian reflector. The 9th-magnitude comet was at an altitude of 39°.*

I mean better than ninth magnitude. Jonathan Shanklin, Director of the BAA Comet Section, lives near the centre of Cambridge, England, a city with a population of about 100,000. He can observe well-placed comets down to ninth magnitude from his home using a 150-mm (6-inch) reflector. But the city comet-hunter must be patient, and be prepared for a long wait before the prey appears.

This dismal picture has been changed completely by the arrival of CCDs. These electronic detectors have a remarkable ability to slice through light pollution, revealing comets that test even country observers (see Chapter 6). Their major drawback for cometary work is the limited field of view of the rather small chips that are currently available. To observe a comet, particularly a large and comparatively close one, you may have to use your CCD not on your main telescope, but at the focus of a fairly fast telephoto lens – say a 200-mm *f*/4, which because of its shorter focal length will give a more compact image than a telescope. The telescope is better for imaging the fainter, newly discovered comets. As with asteroids, CCDs have made possible accurate positional measurements, which are particularly welcome when a comet has recently been discovered.

Visual estimates of a comet's brightness are useful, but city conditions will probably result in inaccurate results, for the reasons given above. Should the comet be bright enough to show a tail, however, another feature to look for is a *disconnection event*, in which a break appears in a comet's gas tail. Comets generally have two tails – a dust tail, which is the one best seen visually, and a gas tail, which points directly away from the Sun. Observations of disconnection events can yield information about the solar wind of particles streaming away from the Sun.

Very, very occasionally there is a comet which throws modesty to the winds and reveals itself in all its glory, tail and all. Comet Bennett in 1970 was one such, and I have yet to see another to match it. It hung like a dagger in the dark morning sky. Shift workers coming off duty were alarmed to see it, as it was poorly publicized and they had no idea what it might be. I could understand why comets were regarded as portents in earlier times. The astronomy publicity machine is much better now than it was in 1970, and if another good comet came along it would attract more media attention, despite the greater light pollution. But comets have let the public down before, most notably Comet Kohoutek in 1973, and of course Halley's Comet in 1985–6, neither of which came up to expectations.

When comets first appear they are usually dim fuzzy patches, indistinguishable in a small telescope from such objects as nebulae and galaxies. While bright comets are visible from city skies, their first faint manifestations are not. Hunting for new, undiscovered comets is a dedicated pursuit and requires good sky conditions. It is most unlikely that a city observer will be the first to spot a run-of-the-mill comet. There is just a chance, however, that a new great comet will appear first in the twilight sky, having crept up unnoticed from behind the Sun. In this case, the first person to spot it could just be some high-rise city astronomer gazing at the distant horizon. Maybe the big one is just around the corner, and the skies of our cities will once more be graced by a spectacular sight that will make everyone, briefly, a skywatcher.

Meteors

Out in the country, the peak period of a meteor shower is a delight to watch. Every few minutes it seems as if a star has broken away from its moorings and dashed with a fiery trail towards the Earth, justifying the popular name of "shooting stars." We know that they are really just tiny flakes of dust from comets' tails that burn up in the atmosphere, but watching for them is still great fun, and can actually be useful. (A point of nomenclature that it is best to make clear: the particle that burns up is the *meteoroid*, the fleeting trail of light we see is the *meteor*. A *meteorite* is a larger, more substantial object that penetrates to the Earth's surface.)

Meteor watching is certainly best suited to those clear, dark skies we all dream of. Simple observations of the brightness and time of meteors, correctly recorded and corrected for observational errors (see page 62), can be of use in charting any variations in the density of the stream of meteoroids which the Earth is encountering. Meteoroids that give rise to a particular shower have their own orbit around the Sun, and the shower occurs at the particular time of year when the Earth travelling in its orbit crosses the orbit of those meteoroids. For some showers the meteoroids are clumped in their orbits, leading to higher rates in some years. The particle size distribution also varies from shower to shower, as revealed by the brightness of meteors observed.

From city skies, however, it is all very different. Instead of a meteor every few minutes, we may be lucky to see one every half hour or even every hour. Dispirited, we give up and curse the handbook that suggested an hourly rate of 60 or 100. The problem, of course, is that the sky is too bright. Because we are rarely looking at exactly the spot where a meteor appears, and because it is a rapidly moving, momentary flash and not a steady source of light, we cannot see meteors anything like as faint as we can stars.

In a fair country sky, with a limiting magnitude of 6.0, we would not expect to see meteors fainter than magnitude 4.5, or maybe 5.0 if we are lucky. Even then, we certainly do not see all the fourth-magnitude meteors that appear in the sky. In a city sky, where the limiting magnitude may be only 4.0 or 4.5, we are restricted to meteors brighter than third magnitude. The numbers of meteors we see usually increases sharply with their faintness: there are actually about two or three times as many fourth-magnitude meteors as third-magnitude ones. So by cutting out a whole magnitude, we are considerably reducing the rate we can record.

Having said that, it is interesting to look at the brightness distribution of meteors actually observed in dark skies. Around 60 percent of the meteors seen in good showers such as the Perseids or Quadrantids are of second magnitude or brighter, and are therefore easily observed even in average city skies. The problem is partly a psychological one – it is hard to sustain the enthusiasm for observing meteors when very few are visible. Even under good skies on the night of a shower's peak activity, there can be long intervals between meteors for a single observer who is viewing the whole sky.

Forecasts of meteor numbers for a particular shower are given as the *zenithal hourly rate*, or ZHR. This is the number of meteors a skilled observer would record under a perfect whole sky of limiting magnitude 6.5 with the radiant at the zenith. The *radiant* is the point from which all the meteors of a particular shower appear to radiate, and it is very rarely in the zenith. If it is an altitude of 45°, only about 70 percent of the ZHR would be recorded – still with perfect sky and perfect observer. If your sky and you are less than perfect, you will see only some of the brighter meteors, and the rate drops further. If you are observing from the city you probably have a restricted horizon, which may cut out another 15 or 20 percent. It is no wonder that you see so few.

Even if you correctly record all the meteors you see during your watch, the results will probably not contribute much to science since so many correction factors have to be applied: for the altitude of the radiant, limiting magnitude and cloud cover. This is less true of what are known as telescopic meteors. As already mentioned, faint meteors are much more numerous than bright ones, and telescopic meteors are those fainter than can easily be seen with the naked eye. You do occasionally see a meteor in the field of view of a telescope, but in fact the instrument most often used for this sort of work is a pair of binoculars.

These need not be anything elaborate – an ordinary pair of 7×50s or 10×50s is perfectly suitable (see Chapter 5). The wider the field the better, but a field of view of about 5° is average for binoculars. The technique is simply to observe a chosen area of sky within about 20° of a meteor radiant. Any meteors that appear can be marked on a sketch map of the area, or details simply noted as for visual observing. One word of caution: because of the limited field of view, rates can be very low, and you may well see only two or three an hour. The good news is that your binocular limiting magnitude is faint enough for you to see virtually as many meteors as a country observer.

Keeping watch on a particular part of the sky using binoculars is physically very tiring unless they are supported. The cheapest method is to buy an adapter which enables you to attach them to a photographic tripod. The mounted binoculars need to be positioned so that you can lie in a garden chair and observe through them at the correct angle. To make life easier, several US manufacturers have devised clever mountings which suspend the binoculars in front of your eyes, with varying degrees of success (*Sky & Telescope*, June 1993, pages 35–40). For telescopic meteor observing you may be able to devise your own version more cheaply, but maybe at the cost of versatility. There is also the possibility, dare I say it, of carrying out this form of observing from indoors (see Chapter 8).

Lights in the sky

Under this catch-all heading are included various natural phenomena of the upper and lower atmosphere, from aurorae to sun pillars, and miscellaneous topics such as observing artificial Earth satellites.

Aurorae

The northern (or, in the southern hemisphere, southern) lights are another phenomenon that many city astronomers just dream about seeing. While in polar latitudes they are a common feature of the long winter nights, for most observers in Britain they are a rarity. The lower the latitude – more correctly, the farther from the magnetic pole – the less likely they are to occur. The north magnetic pole is located in northern Canada, so observers in Canada and the northern US are much better placed for seeing aurorae than their counterparts in the densely populated parts of Europe.

Faint aurorae occur fairly frequently low in the northern skies of Britain and the northern US, but they often go unnoticed since a perfectly dark sky and clear horizon are needed to see them. Although the predominant colour of an aurora is generally green, from lower latitudes we see only the upper part of the display, which more often than not is red. From time to time there is an aurora bright enough even for observers near cities to be aware of it (see Figure 4.10). The sky can become very red, and various distinct patterns, given names such as rays and bands, can develop. The event is better seen from dark skies, but always bear in mind that an aurora may appear at any time. Every time you go out to observe, glance at your northern horizon. You will get to know how bright it normally is, and should be able to detect any change.

If an aurora is in progress, it may last only for a few more minutes, or it may continue for hours. You could jump in the car and head for your darkest local site, but by the time you get there the display could well be over. Instead, it may be worth trying to photograph the event. The eye is not very sensitive to the red light of aurorae, but colour film is. If you have a fast film in the camera, try time exposures of 10–20 seconds or so at full aperture. You will need to keep the camera steady, preferably mounted on a tripod, and it is better to use a wide-angle lens if you have one. You may be lucky enough to record more than you can actually see. (There is more on sky photography from the city in Chapter 6.)

Noctilucent clouds

Aurorae and meteors are both phenomena of the upper atmosphere. A third, less well known but very beautiful and definitely visible from some cities at the right time, is noctilucent clouds. These fine, wispy clouds appear in the summer months in the northern sky. They are more common at latitudes between 55 and 60°, but from time to time they are seen as far down as latitude 50°. This makes them a feature of northern European and

Figure 4.10 *Occasionally aurorae are bright enough to be seen in suburban skies. This one, in March 1989, photographed from the author's home, was strong enough to overcome not only streetlights but a first-quarter Moon.*

Canadian skies, rather than American skies. You may think that fine wispy clouds are nothing unusual, but noctilucent clouds become visible long after the Sun has set. They have a characteristic pale blue colour that is brought out well on colour film, as in Figure 4.11. Noctilucent clouds are quite bright and are visible in the twilight sky, before light pollution has got much of a grip, so they can be seen quite well from cities.

Noctilucent clouds are becoming more frequent, for reasons which are not understood. They form when water vapour condenses on particles at the low temperatures that prevail at altitudes of around 82 km (51 miles). It is not certain what these particles are, but it may be that industrial pollution or the increase in air transport has provided more high-altitude particles around which these clouds can form. Anyone in the right part of the world should keep an eye on the northern horizon after twilight during the summer months for these delicate and beautiful phenomena.

Figure 4.11 *Noctilucent clouds are delicate clouds which form high in the Earth's atmosphere, at five times the altitude of the highest normal clouds. They are visible only from certain latitudes, but are bright enough to be seen in city skies. These were photographed from near the author's home in June 1993.*

The zodiacal light

While on the subject of occasional lights in the sky, I must mention the zodiacal light, if only for the sake of completeness. This is a pale cone of light that appears after sunset and before sunrise along the line of the ecliptic, the path followed by the Sun, Moon, and planets across the sky. It is caused by dust particles way out in space, in the inner Solar System. The best time to see it is when the ecliptic is at its steepest angle to the horizon, which is in spring in the evening sky and in autumn in the morning sky.

You need good, dark, clear skies to see it, as it is chased away by the slightest hint of light pollution. When you do see it from a good location, however, it seems so obvious that you wonder why you missed it before. The reason is that the light is always there, but it is so smeared out by the presence of water vapour and haze that even in the absence of light pollution it is indistinguishable from twilight. Add artificial skyglow and it disappears completely. The zodiacal light is one of those phenomena that city observers must resign themselves to missing.

Solar phenomena

Although not strictly astronomical, various atmospheric effects are of interest to astronomers. These include mock suns, haloes, and sun pillars. A *mock sun* (alternatively called a *sundog* or *parhelion*) is a bright patch of light, often brightly coloured, at the same distance from the horizon as the

Figure 4.12 *This sun pillar was photographed one wintry morning, though the phenomenon can also be seen in summer.*

real Sun but 22° away from it to one side. Quite often you can see two at once, on either side of the Sun. These are most often visible when the Sun is fairly low in the sky, but a solar *halo* can sometimes be seen when the Sun is high up. This takes the form of a pale ring around the Sun, with a radius of, again, 22°. A *sun pillar* (see Figure 4.12) is a bright column above or below the Sun as it rises. All these phenomena are caused by ice crystals suspended in the atmosphere. This does not mean to say that they are restricted to cold weather; at high levels, typically where cirrus clouds form, the temperature is well below freezing even in the tropics.

Another atmospheric phenomenon connected with the Sun is the *green flash* – a green upper rim to the Sun's disk briefly visible just before it sets if the atmosphere is particularly stable. If the setting Sun has a jagged edge as a result of layering in the atmosphere, there is quite likely to be a green flash as long as there is not too much haze. It is best seen through binoculars – but use them only if the Sun is not painful to look at with the naked eye. The green flash is caused by refraction of sunlight by the Earth's atmosphere. A clear, low horizon is needed to see it, so this is a sight for those in high-rise blocks.

Identifiable flying objects

Every so often astronomers are called upon to interpret observations of what are generally termed unidentified flying objects, or UFOs. Curiously, I have never met a regular watcher of the sky who has seen any object which was not readily identifiable. The non-astronomical public are largely unaware of the range of phenomena that can be seen in the sky, and this unfamiliarity gives rise to many bizarre reports which turn out to be easily explained. It is always worth knowing the kind of things that give rise to UFO reports.

Venus, Jupiter, or Sirius when low in the evening sky account for many UFOs, while weather balloons or aircraft are behind many more. Living not far from London's Heathrow Airport, I am familiar with the sight of a long string of glittering beads hanging in the sky – the headlights of several aircraft lined up on the approach path. These can be seen from up to 50 km (30 miles) away. When the wind is coming from an unusual direction, the planes use a different runway and the string of lights is seen by people who are not familiar with the sight.

Another form of UFO, reported particularly from cities, can sometimes be seen with the naked eye or in binoculars. It consists of a pale yellow or orange blob, making its way steadily across the sky with an undulating motion. This is nothing more than a bird with a white belly, illuminated from below by the streetlights.

Occasionally even experienced observers get caught out by blobs of light which seem to dance around the sky. These are caused by banks of searchlights, of the type that often accompany an event such as a rock concert or funfair, playing on a thin layer of haze. Events like this are sometimes held in out-of-the-way places, so the lights could appear where you least expect them. Laser light shows produce starlike points which can suddenly dash across the sky.

Artificial satellites

Many people are surprised to be told that artificial Earth satellites are easy to see, and that their sighting of a strange bright light silently crossing the sky is of just such an object. Sometimes you may be misled by reports that the object moved irregularly, but this is an optical illusion. When the only reference points are stars, people can become disoriented and believe that the satellite, whose motion is essentially straight and regular, is winding its way among the stars or moving in jerks.

While again not strictly astronomical, the techniques of observing satellites are ideally suited to amateur astronomers as the main requirements are a pair of binoculars and a knowledge of the sky. Armed with a set of predictions of the track of a satellite, the observer locates it and makes notes of its exact path. This is done by recording which stars it passes between or close to at a particular precise time. These records can then be converted to RA and dec positions.

It may seem strange that observations of this sort can be of any use at all when satellites are routinely tracked by national space agencies, but they do yield useful information about the decay of satellite orbits as a result of the drag of the upper atmosphere. Most satellites are not tracked constantly, so amateur observations are welcomed by scientists studying the upper atmosphere. The only organization that routinely trains satellite observers and coordinates their work is the British Astronomical Association, which forwards results to the Royal Greenwich Observatory.

Variable stars

If such phenomena as the zodiacal light are beyond the reach of city-dwellers, variable stars are highly accessible. What is more, estimates of variables are among the most scientifically useful observations the amateur can make, and very little equipment is needed.

Most stars are of fixed brightness, but a significant number do vary. With many the changes can be monitored by making careful estimates, either with the naked eye or using a telescope or binoculars. While these estimates may lack the precision of measurements made using electronic devices such as photoelectric photometers (see Chapter 6), they have the merit of being easily and quickly made. One of the advantages of variable-star work is that you can make estimates even when only a small part of the sky is clear. While planetary observers need a good view of the ecliptic, there are variables all over the sky. Furthermore, an estimate can be made in a few moments if necessary, making the most of those nights when it may be crystal clear between numerous scudding clouds.

Variable stars come in all varieties. Some vary because one member of a binary system periodically hides the other from our view, causing a dip in their combined brightness. These are called eclipsing binaries. Pulsating stars, such as Cepheids and RR Lyrae stars, vary as their surface area changes. All these categories of variable star have precise periods of variation. Others vary fairly regularly, such as Mira in the constellation of Cetus, which reaches a maximum brightness usually of about magnitude 3 or 4 every 330 days or so, then beginning its fade to about magnitude 9 (see Figure 4.13). Other stars are semiregular, or may be unpredictably irregular in their variations. There are those that suddenly flare or suddenly fade, and those that suddenly cease their variations for a while. Novae and supernovae undergo huge outbursts which increase their brightness by ten to twenty magnitudes before fading back to obscurity. When one appears, its brightness changes need to be followed closely. With such a motley collection of stars, there is always something happening and individual observers have a real chance of spotting a new development ahead of anybody else.

Figure 4.13 *Light pollution does not prevent you from following the light variations of a variable star using binoculars. These two photographs of Mira (Omicron Ceti) were taken two months apart; the star has brightened from about magnitude 7 in (a) to magnitude 3 in (b). It can vary between magnitudes 2 and 10.*

(a)

(b)

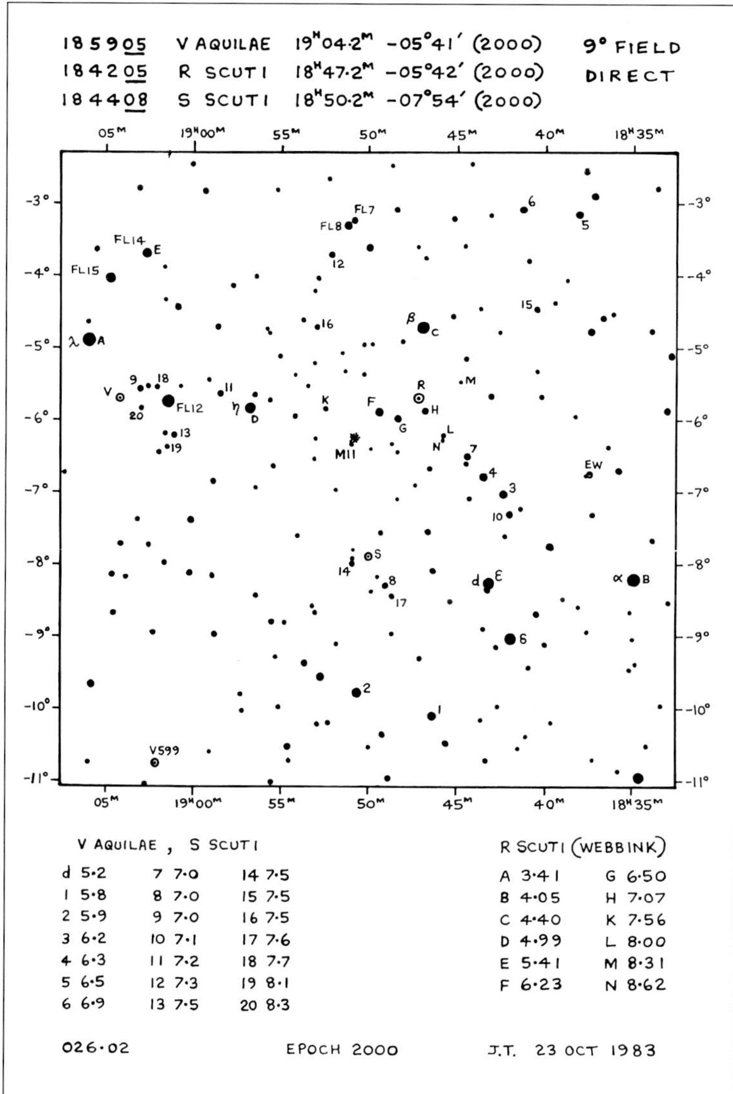

Figure 4.14 *A variable-star chart supplied by the British Astronomical Association's Variable Star Section showing the field of the variable stars R Scuti, S Scuti, and V Aquilae, and the magnitudes of comparison stars. The stars in the Scutum area are easily visible with the aid of binoculars from anywhere in the world, from February (in the morning sky) to November (in the evening sky).*

There are several techniques for making variable-star estimates, but basically you compare the brightness of the variable with that of other stars of known, fixed brightness. These comparison stars should be fairly close in the sky to the variable, ideally within the same field of view of the instrument you are using. Some should be slightly brighter than the variable, and some fainter. If the variable has a wide range of brightness, then many comparison stars are needed to cover the whole of its range. Another requirement for comparison stars is that they should not be strongly coloured, since the eye's sensitivity to colour varies with brightness.

Good lists of comparison stars for variables are available from organizations such as the American Association of Variable Star Observers (AAVSO) and the BAA Variable Star Section. These lists usually take the form of detailed star charts of the region of the variable, with the comparison stars indicated. An example is shown in Figure 4.14. Some charts mark the comparison stars by letters, and a separate list gives the magnitudes of the comparisons; on others the magnitudes are marked directly to the nearest tenth (but with the decimal point omitted, so that 58, for example, indicates magnitude 5.8). Armed with these, you first locate the area of the variable by using a more general star atlas. Then comes the business of making the estimate itself.

The basic method is to choose two comparison stars which are slightly brighter and fainter than the variable, and then to work out where the variable fits between them. There are two variants. If you are using the BAA's *fractional method*, you make an indirect estimate. If the variable (V) is just slightly fainter than star A and quite a bit brighter than star B, for example, you may decide that it is a quarter of the way from A to B. You would then record your estimate as A(1)V(3)B. Back indoors, you convert this into an actual magnitude. With the *AAVSO method*, you make a direct magnitude estimate. Knowing your chosen comparison stars to be magnitudes 7.5 and 8.1, say, you picture the difference between them as six 0.1-magnitude steps, and place the variable on this scale. With both methods, the process should be repeated with several pairs of comparison stars, and the results averaged. Most other national observing organizations have adopted one of these two methods.

The human eye is actually quite sensitive to small differences in brightness, and once you have a little experience of making estimates, you can begin to train yourself to recognize magnitude differences using just one comparison star. Then you can use *Pogson's step method*, and estimate the variable's brightness more directly. It may appear 0.4 magnitude brighter than one comparison star, say, so you have an immediate estimate of the variable's brightness. You then repeat the exercise with another comparison, and average the results. If necessary, it is possible by this method to make an estimate using just one comparison star.

The beauty of variable-star observing for the city observer is that, even from a badly light-polluted location, quite faint stars are visible with a minimum of optical aid. The background sky glow may be bright, but its effects can be reduced by using more magnification. John Isles, former Director of the BAA Variable Star Section, recommends using slightly higher-power binoculars from the city than you would otherwise choose. He favours 12×40 binoculars for this purpose, though the more common 10×50s are also suitable, and even lightweight 8×30s for the brighter variable stars, those of magnitude 4 to 6. Although these stars may be visible with the naked eye even from the city, binoculars will make them much more easily observable and will therefore improve the accuracy of the estimates.

John made up to 2000 variable-star estimates in one year using such binoculars from the roof of a building in London's Covent Garden. Those unfamiliar with London might think from the name that this is an oasis of greenery, but in fact it is in the heart of theatreland and has one of the most light-polluted skies you could imagine. He advises that, as with telescopic meteor observing, it is a good idea to mount the binoculars on a tripod. This is especially useful if you want to see stars near the limit of your sky, as city observers are very likely to do. With 12×40s he could reach magnitude 9 on a good night, and magnitude 8 on an average night.

One word of caution John offers to city observers of variable stars is that if your sky is noticeably orange as a result of sodium lighting, red variables such as Mira type stars and some semiregular variables will tend to appear fainter than they really are. It is then best to concentrate on other types of variable. You may be tempted to try to suppress the glow using light-pollution reduction filters (see Chapter 6), but these are frowned on by serious variable-star observers as they may give rise to systematic errors which can vary from star to star. This is because there may be strong features in the spectrum of the star which some filters would cut out. However, the difference is likely to be small, particularly when you are dealing with a star which has a large range of variability.

Variable-star work can also be carried out by telescope, to reach fainter stars than are visible through binoculars. Among the objects that can be monitored in this way are active galactic nuclei. These are the centres of galaxies in which a huge amount of energy is being generated from a very small region, and this energy output can vary irregularly. In many cases the galaxy itself cannot be seen, and certainly not from the city, but the nucleus itself can be located in just the same way as a variable star. Professional astronomers regularly ask the amateur community to keep a particular object under surveillance, and charts are issued giving suitable comparison stars. Such objects are usually fainter than 12th magnitude, so a telescope of fairly large aperture is needed.

Some enthusiastic amateurs are now using photoelectric photometers, either bought off the shelf or custom-built. These instruments are suited to making highly accurate measurements of stars which vary by a few tenths or even hundredths of a magnitude, which is beyond the capabilities of the human observer. This area is covered more fully in Chapter 6.

The army of amateur variable-star observers is mobile, widespread, and versatile, and plays an important part in this field of astrophysical research. It is a field where cooperation is very important, and it is more or less essential to join a group such as the American Association of Variable Star Observers or the BAA Variable Star Section, so that you can use the same charts and comparison stars as others.

Nova and supernova patrols

Novae and supernovae are stars which suddenly increase dramatically in brightness. Most observed novae are within our own Galaxy. They undergo an increase of typically 10 to 15 magnitudes, which raises them from the ranks of remote and undistinguished stars in the Milky Way to naked-eye magnitude. Supernovae undergo much more violent outbursts, maybe a staggering 20 magnitudes, which means that for a few days they may outshine all other stars in a galaxy. Supernovae are usually seen in distant galaxies, but there is no reason why one should not appear in our own, except that they are exceedingly rare. The last to be observed in our Galaxy was in 1604, though observers in the southern hemisphere were treated to a supernova in the Large Magellanic Cloud, a small companion galaxy to our own, in 1987. This reached third magnitude, despite the fact that it was about 170,000 light years away. Had it been within our own Galaxy and in a location unobscured by intervening material, it could easily have overtaken Sirius as the brightest star in the sky.

Some dedicated amateurs search for novae and supernovae. Nova-hunters scan the Milky Way, where most novae occur, looking for stars that are not usually there. This can be done visually, which may seem an impossible task, but some people really do know the sky that well. If the idea of learning the position of every star down to ninth magnitude seems daunting, you could always start by getting to know a selected area, then expanding the search. George Alcock, who has discovered six novae and five comets (see page 160), has a photographic knowledge of the sky and can spot interlopers almost as soon as he sees them. Similarly, American observer Peter Collins, discoverer of three novae, has spent thousands of hours learning the stars of the Milky Way down to eighth magnitude. But be warned – the time it takes to find just one nova or comet can be 1000 hours or more, even in good skies.

However, there is a method which is less demanding on your memory: the photographic patrol. You take photographs of the same regions of the Milky Way night after night, and compare each with a master photograph. One way to do this is to make a *comparator*, which at its simplest consists of two magnifiers of equal power. You look through these with both eyes at the two images, and align them so that the star images merge. Any star that has changed in brightness, or appeared where there was none before, will show up as an apparent ghost image in one eye. Another way to do this is to use the PROBLICOM® system pioneered by Ben Mayer in the US and put to good use by William Liller. This uses two identical slide projectors to align the images on a screen, switching rapidly from one to the other. With this arrangement any differences appear to blink on the screen, giving the system its name, which comes from "projection blink comparator."

The photographic method seems simple in principle, but there are practical difficulties. Liller now lives in Chile, where the skies are darker and more uniformly clear than in most other places. The comparison method works best where the two images are identical apart from any varying stars, and in urban areas varying levels of light pollution can produce marked variations in the sky background brightness. The other problem is that there are plenty of other objects which can produce additional star images. Known variable stars and asteroids are real objects which can be identified and eliminated, while photographic flaws and ghost images of bright stars are more difficult to discount. Rotating Earth satellites that flash briefly as they glint in the sunlight can also produce spurious images. One way to overcome these problems is to make two identical exposures, shifted by a uniform amount, on the same frame of film. Flaws are unlikely to have the same double appearance, while ghost images are generally to be found diagonally opposite the bright star in the frame, so they will shift position between the two exposures.

There is no reason why a city-dweller should not discover a nova. Most discoveries are made when the objects are brighter than ninth magnitude, quite bright enough to be seen from the city. What counts more than dark skies is dedication and enthusiasm – and perhaps the desire to make a genuine discovery. In some tourist traps you can buy newspapers with headlines which go something like "(Your name here) is the richest person in world!" Imagine the official announcement flashed to the observatories of the world which says "(Your name here) has reported the discovery of a nova ..."!

Supernova patrolling has a similar appeal. First you locate a galaxy, then you compare each of the nearby stars with charts or photographs of the area, looking for newcomers. Many visual discoveries of supernovae have been made in this way by the Reverend Robert Evans, observing from country locations in Australia. From the city, the job is made that much harder if you cannot see the galaxy in the first place. The supernovae themselves are generally faint – 13th magnitude is considered bright – but objects of this brightness are within the reach of many amateur telescopes, even in city skies. Charts are available which show the field stars around a selection of likely galaxies.

A computer-controlled telescope is a major advantage in this sort of work, since it cuts down the time needed to locate each galaxy. With the addition of a CCD imaging system, it should be possible to discover supernovae from urban locations. Professional astronomers at Berkeley, California, have been operating a CCD system for some time. Their prototype was plagued with problems, and it took seven years of experiment and effort before the first supernova was discovered. Today they achieve regular successes. With the advent of off-the-shelf computer controllers, CCDs, and software, much of the development work has been done for you.

Double stars

Individual stars, on the whole, are rather monotonous. But some otherwise ordinary stars are real beauties because they form part of a double-star system. They are among the showpieces of the sky, and feature high on the city astronomer's list of favourite objects. While some double stars look close just because they happen to lie in the same line of sight from here on Earth, most are physically associated binary systems – two stars in orbit around each other.

There can be few August star parties where someone is not gazing at Albireo, Beta Cygni (see Figure 4.15). You can have endless discussions about whether its two components are blue and yellow, green and yellow or just plain white. Even plain white doubles can be attractive to look at, notably where there is more than a straightforward pair of stars to see, such as in the famous "Double Double," Epsilon Lyrae. There are many other popular doubles in the sky whose appeal lies in their beauty. But there is a serious side to double stars if you want to get more deeply involved.

Measurements of double stars, like those of variable stars, are scientifically useful. The fact that double star work is unpopular with amateurs shows that scientific usefulness not surprisingly takes back place to other considerations. Most amateur astronomers do what they do for fun, and any worthy results are byproducts. To measure double stars you need a telescope with a reasonably long focal length and good optics equipped with a *micrometer* (see Figure 4.16). With the micrometer you can measure the *separation* and *position angle* – the relative orientation – of the two stars. The device is fitted in place of the eyepiece.

Figure 4.15 *Albireo, or Beta Cygni, is a star which can be seen as a double with any power greater than about ×10. Observers disagree over the colours of the stars, which are emphasized by the contrast between them, but this photograph shows them as blue and yellow.*

Looking through a typical filar micrometer, you see a cross-wire superimposed on the field of view, and a movable third wire parallel to one of the wires of the cross, as in Figure 4.17. The whole assembly can be rotated through 360°, and either the field or the cross-wires can be illuminated. To use it, you centre one component of the double star on

Figure 4.16 *A filar micrometer fits in place of the normal eyepiece and has two vernier scales for measuring the separation and position angle of double stars, or the positions of planetary features.*

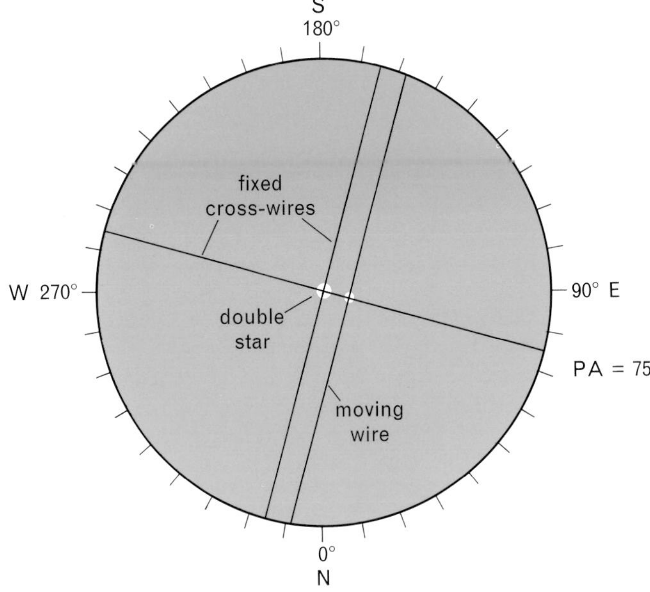

Figure 4.17 *When using a filar micrometer to measure a double star, first align the double along one wire, with the main component of the double exactly on the fixed cross-wire. Then bring the movable wire across to intersect the other star.*

the cross-wire, with the companion aligned along the arm of the cross-wire which is crossed by the movable wire. You then bring the movable wire up to intersect the companion, so that both stars are accurately positioned at each of the two points at which the wires cross. The separation of the wires allows you to measure the separation of the double, and the orientation of the assembly gives the position angle. The micrometer must be accurately calibrated so that you can convert the micrometer settings into actual measurements.

The lists of double stars which appear in practical observing handbooks, such as *Norton's 2000.0* star atlas and *Burnham's Celestial Handbook*, give these figures, so it might seem that there is little more to be done. But double stars are very important in astrophysics, because they provide a direct way of discovering the masses of individual stars. Since in many binary systems the two stars appear to move around each other very slowly, continued measurements help to refine the orbit and perhaps reveal other companions. The sky is full of double stars, so there is plenty to be done. And, like other stars, they are not seriously affected by skyglow. This is work that can done from the city just as well as from a dark country site. It does, however, demand an equatorially mounted telescope with a focal length long enough to make it easy to obtain high powers.

According to double-star observer Bob Argyle, of the Royal Greenwich Observatory, the minimum aperture you need is about 200 mm (8 inches). He observes from a site about 1.5 km (1 mile) from the centre of Cambridge, only 300 metres from a main road lit by high-pressure sodium lights. He finds that with a 200-mm refractor he can make double-star measurements nearly every clear night from this location, and has made more than 10,000 in three years.

Quite faint double stars are often neglected, though ones whose components are between 2 and 10 arc seconds apart are easy to separate with even a small telescope. But a moderately large aperture is needed to give good images of the stars rather than just separate them, and Bob regularly uses a magnification of 450 from Cambridge. He emphasizes the need for a solidly mounted telescope with a good drive and slow motions: "At high magnification the slightest bump can move the images considerably and if, in trying to set the wires you need both hands to continuously move the RA and dec slow motions, then ...." He leaves the picture of frustration to the imagination. This requirement probably rules out the commercial Schmidt–Cassegrain telescopes on their standard mountings.

Micrometers are available commercially, though at rather high cost – about the same as a cheap camcorder. This is not the sort of work that people do on a casual basis, so anyone who wants to contribute to knowledge in this way must justify the cost in terms of their own satisfaction. The small number of active amateur double-star observers around the world do find, however, that the work they do is almost certainly not being duplicated by others.

Amateur astronomers have used CCDs to improve their observations in virtually every other field of observing, so what about double stars? A fairly large image scale is needed – at least 0.5 arc second per pixel (picture element), and the exposures should be made through filters to reduce the tendency for the Earth's atmosphere to act as a prism and smear a star image out into a short rainbow. The technique used by professionals is first to find the orientation of the CCD on the sky by trailing star images across the field. Then well-known doubles are observed to establish the image scale. Software currently available to amateurs would have to be adapted to produce the actual measurements. This is a field ready for exploration by amateurs.

Deep-sky objects

The technique of observing deep-sky objects – that is, clusters, nebulae, and galaxies – has changed radically over the past 20 years or so with the introduction of filters designed to cut out light pollution. These are covered in Chapter 6; here I shall deal with observations from the city of the various types of object made using binoculars or a telescope with no filter.

The lessons of Chapter 2 – wait for the clearest nights and aim high – must be heeded when you are trying to find deep-sky objects from the city. Make sure that there are no lights directly in your line of sight. One neat trick is to block out such lights by hanging a blanket on a clothes line. Some people make screens which they set up when they are observing to shield particularly troublesome lights. The section in Chapter 5 on maximizing contrast within your telescope is also essential reading for the new deep-sky observer. Good baffling inside the telescope can make an enormous difference to your ability to see low-contrast objects.

Don Miles of the Webb Society – an organization devoted to perpetuating the pioneering deep-sky observing work of the Reverend T.R. Webb in the nineteenth century – observes with a 200-mm (8-inch) Schmidt–

Cassegrain telescope from Lovedean, which despite its idyllic name is actually on the outskirts of the crowded, urbanized island of Portsmouth on the south coast of England. He has the following tips.

Eyepieces in particular get dirty easily, and Don cleans his at least every other observing session. He uses Kodak lens cleaner, which is specially designed for use on coated optics. He claims that he can add up to a magnitude in his light-polluted sky by keeping the optics clean. One advantage of a Schmidt–Cassegrain or a windowed reflector is that the main mirror is not exposed to the atmosphere, so it remains in good condition for longer. The exposed glass, however, is very prone to collecting dew and Don uses a dew cap which projects beyond the end of the instrument a distance of one and a half times the aperture. Not only does this prevent dew from forming, it also keeps stray light off the corrector plate.

Don advises, "Don't think because the sky's bright you don't need to dark adapt. If it's not darker outside than in your lounge, you should pack up anyway!" There are several methods of preserving your dark adaption. One of your most useful accessories can be a piece of black cloth. Use it to cover your head and eyepiece while you are observing, like an old-fashioned photographer. The total darkness keeps your eyes fully dark adapted and helps you to catch faint detail in the field of view. The long-term improvement in dark adaption may even make you think that the sky has got worse after an hour or so of observing, whereas all that has happened is that your eyes are better adapted.

Alternatively, wear an eyepatch over your non-observing eye. It is always better to observe with both eyes open, to avoid the discomfort of keeping one eye closed. You may even go to the extreme of using one eye for looking at your star charts and the other for observing, using eye-patches to alternate. This is similar to the trick used by Second World War fighter pilots who would close one eye when flying into the Sun so as to preserve the vision of their other eye should danger then threaten from a different direction.

Rubber eyeshields on your eyepieces also help to exclude stray light. One very cheap and simple alternative has been suggested by Michael B. Leigh of Lakeside, California. He buys black terry-cloth wristbands which he slips over the end of the eyepiece. They also reduce the risk of "image shake," which would happen if you bumped the telescope by getting too close. And if you leave the telescope for a while, he points out, you can fold the band over the eye lens to protect it from dew.

While the country observer can move from one deep-sky object to another according to whim, the town-dweller must plan carefully. Select objects that will be as near the zenith as possible, and take note of catalogue listings that indicate how condensed they are. The more condensed objects are likely to be more visible in city skies, even faint ones. Some catalogues list surface brightness, which is helpful though not infallible. One which does is the *Observing Handbook and Catalogue of Deep-Sky Objects* by Christian B. Luginbuhl and Brian A. Skiff – a constellation-by-constellation guide to some 1500 objects for medium-size telescopes.

A brief word for the uninitiated about catalogue numbers assigned to deep-sky objects. The ones you will encounter most often are those prefixed by "M," "NGC," or, to a lesser extent, "IC." Those with M numbers were originally listed by the French comet-hunter Charles Messier in the eighteenth century (there were a few later additions to his list), and are now referred to as *Messier objects*. Messier's aim was simply to note, for his own use, those objects that could be mistaken for comets. He used telescopes which were effectively about as good as a modern 90-mm (3½-inch) refractor. You might imagine that his list would include the brightest objects observable from the latitude of Paris. However, the skies of Paris were then darker than those of modern suburbs, and many Messier objects are now hard to see, except from good locations.

One of the most successful nebula-hunters was Sir William Herschel, who used a telescope with a 470-mm (18.7-inch) metal mirror, with the light-gathering power of a modern 330-mm (13-inch) telescope. The observations from which his catalogue was compiled were made from Slough, to the west of London, and the list was extended by his son John to include southern objects. It was to become the basis of the *New General Catalogue of Nebulae and Clusters of Stars* prepared by J.L.E. Dreyer of Armagh Observatory. Dreyer's numbers are those we use today, prefixed with NGC. Later additions he included in two *Index Catalogues*, and numbers from them are prefixed with IC. The descriptions and positions have been brought up to date in *NGC 2000.0*.

Clusters

A cluster of stars, whether a loose *open cluster* or a compact *globular cluster*, is a fine sight in a dark sky with a good wide-field eyepiece. Sadly, the city observer has to be content with less spectacular views. Although it is true that increasing the magnification darkens the sky background while leaving star images the same brightness, this also has the effect of reducing the field of view. People pay a lot of money these days for field of view. The new designs of eyepiece that cost and weigh as much as a telephoto lens for a camera can give superb definition over an enormous field of view compared with ordinary orthoscopics and Kellners. You almost feel you could climb inside and be there. They are designed in particular to give wide fields with instruments of *f*/5 or shorter. Other, less costly eyepieces can be used on instruments with longer *f*-ratios. (Eyepieces are described in Chapter 5.)

Any solution that reduces the field of view is not very welcome when the whole point of looking at the object is to gasp with amazement. The city observer must forgo seeing a thousand pinpricks of light against a velvet cushion of sky. But the ultra-wide-field eyepieces do seem to offer the city observer the advantage of their improved contrast. David Cortner of Johnson City, Tennessee, has skies with a naked-eye limiting magnitude of about 4.5 which barely show the Milky Way, even when it is overhead in summer. He has this to say about eyepieces:

> Every time I do a side-by-side comparison, Tele Vue's eyepieces really do seem to deliver an appreciable boost in contrast, and that's what's most lacking in the city sky. My Naglers, a WideField, and a Plössl have made a huge difference in what I can see in many favorite objects (M51 is a vaguely dynamic shape in my old 20-mm Erfle; in my 16-mm Nagler it is a hurricane made of soft, milky light,

shot through by bright and dark detail). The fact that both my primary instruments are short-focus (*f*/6 and *f*/5) may mean that the pricey glass has more benefits for me than it might offer for someone using long-focus optics.

Howard Brown-Greaves of London found that his 13-mm Nagler eyepiece performs better than his 16-mm Nagler, which has an extra glass element. The 13-mm gives better contrast, partly because its lenses are fully coated, but also, he believes, because the extra field of view helps to make a faint object stand out against the background sky. The additional magnification which you can use while still retaining a wide field of view may also be a help.

Even with a basic eyepiece, such objects as the Double Cluster in Perseus, the globular cluster M13 in Hercules, and M35 in Gemini are worthwhile targets from the city. These clusters contain bright stars which should be easily resolved in a moderate instrument such as a 150-mm (6-inch) reflector. But they are by no means the limit of what you can see. Try looking for the fainter globulars and the smaller loose clusters which force you to use a higher power. Even brief listings, such as those in *Norton's 2000.0*, include a selection of non-Messier clusters which can be seen with no difficulty.

With the fainter objects, good seeing may well play a part in turning them from a vague blur into something worth looking at. Many globular clusters should be visible from the city, but you will probably not see as much of them as from a dark site.

Although you may well be able to see faint globular clusters which are well placed, some bright Messier objects can be hard to see. From Britain I have never been able to locate the most southerly Messier object, M7 in Scorpius, even in a comparatively dark sky, because it is always less than 5° above the horizon. I know people who have seen it, but I have never been lucky. Yet from a site farther south, where it rises higher in the sky, it is a very prominent and beautiful cluster, and a spectacular sight in binoculars. When I see it from a location such as Tenerife, I wonder how I could ever miss it from home. This shows the importance of making sure that the object is well placed before you try to observe it.

Many people attempt to see all the Messier objects. This is not a difficult task provided you have the right site, but from a city – particularly in Britain – it is a daunting challenge. The fact is that the list of Messier objects is not a "top 100" of the brightest nebulae and clusters in the sky. It includes many obvious clusters, such as the Pleiades (M45) in Taurus, shown in Figure 4.18(a), and the Wild Duck (M11) in Scutum, but fails to include the Double Cluster (NGC 869 and 884) in Perseus, one of the showpieces of the sky, and of course most southern-hemisphere clusters such as the splendid NGC 3532 in Carina. Rather than try to catch a glimpse of the fainter Messier objects, which are elusive and unrewarding targets from the city, it would be better to look for brighter non-Messier objects which are actually much more interesting.

You will find many beautiful sights in the sky which are not included in any of the main listings. Plenty of groupings of stars visible in low powers (see Figure 4.18) do not appear in the major catalogues. To observers in earlier days, who did not have to put up with city lights and had much else

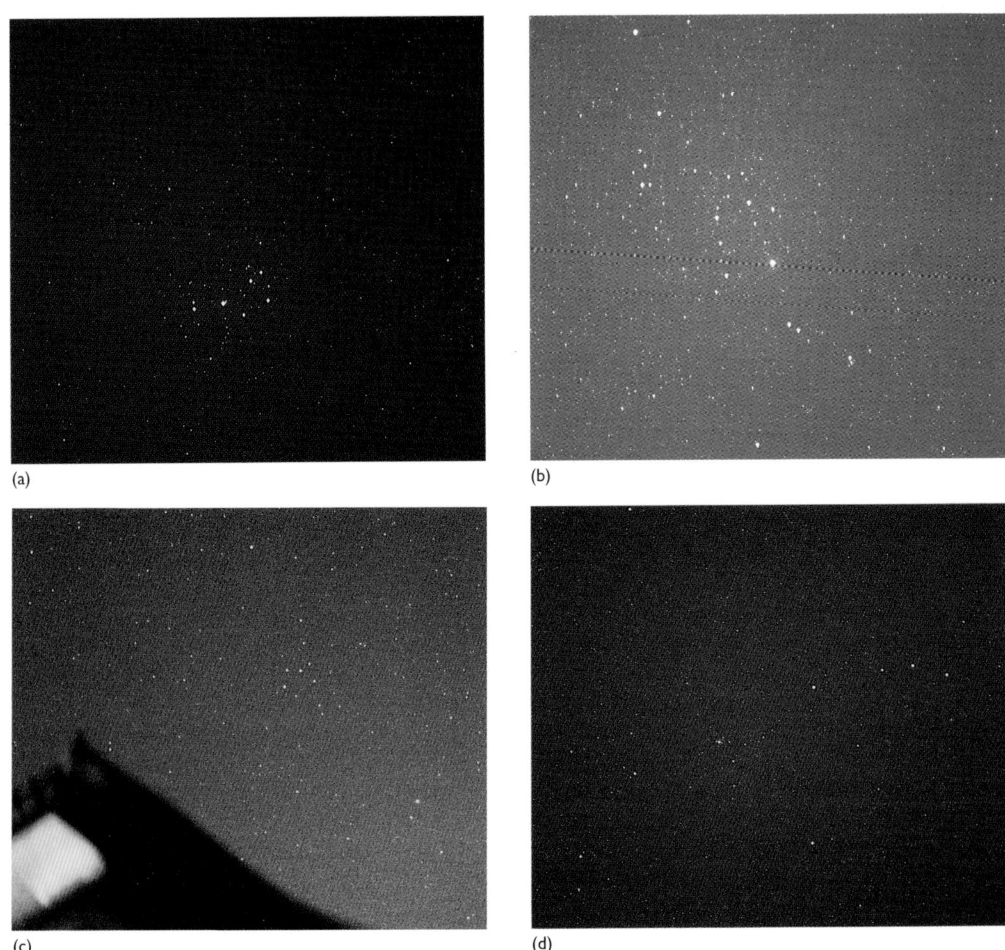

Figure 4.18 *Star clusters can be observed easily from city locations. (a) The Pleiades (M45), the most prominent of all clusters and a splendid sight in binoculars. (b) One of the unsung clusters in the sky, the Alpha Persei Association. Best seen in binoculars, its stars are more widely spaced than those of the Pleiades. (c) At the upper left of this picture lies the Coathanger, a group of 10 stars of 6th and 7th magnitude in a distinctive shape. (d) Kemble's Cascade is a line of stars best seen with a power of about ×50. The cluster NGC 1502 lies at its lower end. Groupings like these last two are not usually labelled as such on star charts, but they are fun to locate and even discover. The Milky Way is a rich area for lines, triangles, and circlets of stars that you can make your own.*

to look at, they did not seem worth mentioning. One example, which I always feel I discovered for myself, is the group of stars around Alpha Persei. In binoculars this is as splendid a sight as the Pleiades seen through a telescope. It is actually known as the Alpha Persei Association, and is a genuine cluster of stars about 570 light years away, according to *Burnham's Celestial Handbook*, that excellent source of heavenly goodies. Another well-known cluster is the Coathanger in Vulpecula, which has the obscure catalogue number Collinder 399. Even *Burnham's* misses this one, yet it is a delight with a low power. It is, however, shown and mentioned in *Norton's 2000.0*.

Walter Scott Houston, writing in *Sky & Telescope* (November 1991, page 559), gave this description of a line of stars he called "Kemble's Cascade": "In this pallid corner of Camelopardalis, at about right ascension 3h 57m, +63° (2000.0 coordinates), was a celestial waterfall of dozens of 9th and 10th-magnitude stars. Down it went, tumbling steeply southwestward over 3° before splashing into the open star cluster NGC 1502." To David Cortner in Tennessee, this string of stars is "as much fun to fall along from town as from out in the dark. NGC 1502 at one end of the cascade is easy and rewarding from inside the city." David Frydman in London gasped when he first saw NGC 457 in Cassiopeia: "It looked just like a praying mantis, with two long lines of stars spreading away from a bright star in the centre." Others call this the Owl Cluster, or even the ET Cluster from its imagined resemblance to Steven Spielberg's long-armed extraterrestrial. All you need to find such objects is the yen to go exploring.

Nebulae and galaxies

There is an enormous difference between a nebula in our own Galaxy and a whole distant galaxy, of course, but to the city observer both present the same difficulty: they are faint, fuzzy blobs. In general, the nebulae cover a larger area of sky and are more diffuse, while galaxies are more regular in shape – usually elliptical – and have a central condensation. This central region is the brightest and easiest part to see, and indeed with most galaxies it is all that is visible. Seen from the city through a moderate aperture, say 150 mm (6 inches), the Andromeda Galaxy is an elliptical blur about the same angular diameter as the Moon, with another object, M32, some distance away. Compare your view with a textbook photograph and you will see that M32 lies against a background of the spiral arms of M31. What you see as M31 in your binoculars or telescope is in fact just the nucleus. In good skies some observers can see the galaxy extending to 5° or more, instead of a miserable half degree. The same applies to many other objects.

Another caution, probably not necessary, is not to expect too much from telescopic views. The vivid colours of galactic nebulae in photographs in books are mostly real, but the eye is just not sensitive enough to see them properly. Many people can see the Orion Nebula as pale green. I must admit that to me it just looks grey, though I often see a reddish tinge to the outer wings when viewing it under good conditions with a moderate aperture. Photographic film, however, can record the colours very easily, particularly the red colour of hydrogen gas (see Figure 4.19).

You might, incidentally, imagine that by getting a larger telescope you will be able to see the Orion Nebula in brilliant colours. In fact, even through the biggest telescopes it has the same surface brightness as it does through small ones. No telescope can make an extended object appear any brighter: all it can do is provide enough magnification for you to see it better. You cannot, for example, construct a one-power telescope that will make everything look the same size but brighter. Similarly, even if you were to be able to travel to the Orion Nebula in a spaceship and look at it from close up, it would still not appear any brighter. It would be bigger, but you would not be able to see the colours significantly better than if you could make it look the same size through

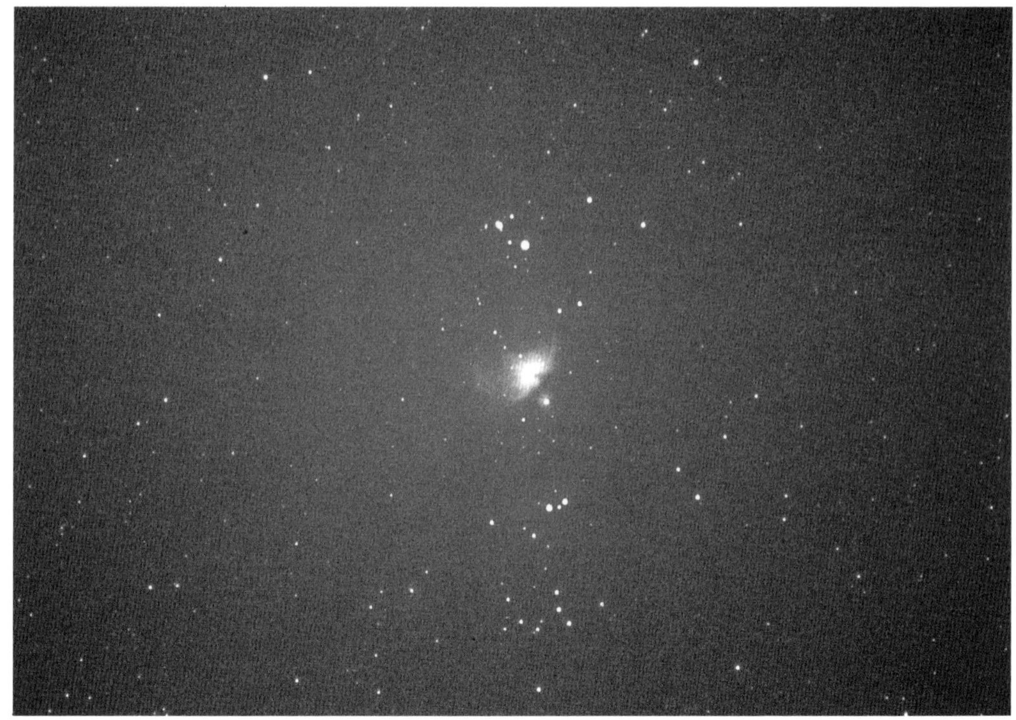

Figure 4.19 *The Orion nebula (M42) is the brightest nebula in the sky and can be observed successfully even from a light-polluted location without using a filter. This photograph, taken using a 400-mm lens, covers approximately the same field of view as a pair of 10×50 binoculars. Visually you will not see the colours, though they are easy to record on film.*

your earthbound telescope. (Actually, from close by you would have less glass in the way to absorb light, no Earth atmosphere, and no interstellar dust, so it would be a little brighter.) If you find this hard to believe, try looking by day at a distant white-painted house or similar object. The white area is just as bright per square degree no matter whether you use the naked eye, low-power binoculars, or a high-power telescope. The total amount of light gathered by a large telescope is greater, but it is spread over a larger area. If the house is too remote to register as a finite area on your retina, you simply will not see it at all. The same applies to deep-sky objects.

As with clusters, the Messier list is not a good guide to the visibility of nebulae and galaxies. Compare, for example, two objects visible towards the end of the year: M33 and M76. The galaxy M33 in Triangulum is supposedly fifth magnitude, yet it is very hard to see in light-polluted skies except for a hint of the nucleus, and in such skies it is hardly worth the bother of searching for it. By contrast, the planetary nebula M76 in Perseus is sometimes described as "the faintest Messier object," and is listed as a pair of 12th-magnitude nebulae in *NGC 2000.0*. Yet it is a comparatively easy object to observe because it is very condensed. Writing in *Sky & Telescope* (November 1993, page 104), Walter Scott Houston suggested that it is visually magnitude 10 or brighter. It is less than 5 arc minutes across, compared with about a degree for M33.

The magnitudes given in observing guides are often photographic magnitudes. To the eye, the objects may appear either brighter or fainter, depending on their colour. What is more, those magnitudes are *integrated magnitudes* – they are a measure of the brightness of the entire object as if it were condensed into a single point of light, instead of being spread over an area of sky. The actual size of the object is far more important. This is particularly true of planetary nebulae, which can vary considerably in size. Apart from anything else, you need to know what sort of eyepiece to use when searching.

Even in perfect skies you need a large telescope to observe the fainter galaxies, and this is doubly true from the city. Aperture is more important than good contrast – a big, cheap Dobsonian is preferable to an expensive refractor. Even so, finding faint nebulae and galaxies from the city can be tricky. It is often fruitless simply to sweep your telescope across the general area of the object in the hope of picking it up. To help, you can buy sets of cards which show the area of the object, including the nearby stars down to a much fainter limit than is shown in star atlases. These can make all the difference to finding an object – they show you exactly where to look and can help you to observe objects as much as one magnitude fainter than you would otherwise be able to find. They are available for selected Messier and NGC objects.

One advantage of having the information on cards is that you can prepare your schedule beforehand, selecting cards for objects that will be high in the sky, and arranging them in order of right ascension so that you know when each will be at its highest. For objects that do not have a card, you may have to resort to the high-tech alternative of making your own guide charts using a computer and the appropriate catalogues on CD-ROM (a compact disc format designed to store data). Those with computer-controlled telescopes, however, can simply punch in the number of the object and let the telescope do the work. For good results, you will have to calibrate the instrument on a nearby bright star just before the observation.

The fact remains that from a light-polluted environment you just cannot see as much as from a dark site. Even so, you can locate a surprising number of objects. In a five-year period from a site only 6 km (4 miles) from central London, David Frydman observed a total of 600 deep-sky objects, using a 125-mm (5-inch) refractor which he says is the equivalent of a good 150-mm (6-inch) reflector. Of these, 200 he saw only once, but the others he observed at least twice.

CHAPTER 5

Choose your weapons

You can be a successful amateur astronomer without owning any form of optical instrument. There is satisfaction to be gained just from naked-eye observing, even with a light-polluted sky. You can strain to catch a glimpse of the Double Cluster, or the Orion Nebula. But it has to be admitted that after a while the excitement of doing this palls, and your thoughts turn to buying a telescope. For the suburban astronomer, as opposed to one living in the country, there is more to consider and perhaps less freedom of choice when it comes to buying equipment. For the best results, the apparatus must suit the observing conditions.

To most people it seems obvious that the one thing an amateur astronomer needs is a telescope. Yet some leading observers rarely use telescopes – instead, they make their discoveries with binoculars. It is now standard advice to beginners that the ideal starting instrument is not a telescope, but a pair of binoculars.

Binoculars

Binoculars have many advantages over a telescope. They are comparatively cheap because they are mass produced, and much better value for money than a very small telescope. They do not need an expensive mount – your arms are good enough for most purposes. They reveal a wide range of objects in the sky that are quite invisible to the naked eye. And you can of course use them in the daytime for a wide variety of other purposes.

I shall give some tips about buying binoculars, since few books go into the matter in any detail. Binoculars are rated by the size of their main lenses – their *objectives* – and their magnification. So a pair of 10×50s have a magnification of ×10 and objectives 50 mm (2 inches) across. It is the size of the objectives that determines how bulky a pair of binoculars really is. A pair of 7×50s is virtually identical to a pair of 10×50s, which have a higher magnification.

The standard size of binoculars for general-purpose observing is usually said to be either 7×50 or 10×50. Of the two, city observers will probably choose 10×50, since the extra magnification will help to give a slightly darker background. The 7×50s come into their own in really dark skies. But there is a place for other sizes, such as 8×30, which are noticeably less heavy. This is important if you observe for any length of time without support, for it is better to have a pair you can use with comfort than a pair you frequently have to put down to rest your arms. I would advise against the more ambitious sizes, such as 20×70, until you are sure you need them.

You need not pay a great deal for binoculars, but you should certainly not go straight for the cheapest. Test a number of different pairs before you buy, including the more expensive models, so you know what is good and what is bad. There are some very high-priced binoculars on the market aimed particularly at the lucrative bird-watching market. Their prices are pushed up by features such as coatings on all the glass surfaces and additional lightness and ruggedness, which are not critical in astronomical use. Birdwatchers have other requirements, such as being able to see a bird against a bright sky, which is where the extra coatings help. A very expensive pair of birdwatching binoculars may not actually perform as well on the night sky as ordinary 10×50s.

Unlike telescopes, binoculars can be tested reasonably well during the day (though stars are always the best test for astronomical purposes). To evaluate a pair of binoculars, choose a test object such as a TV aerial that stands out against a bright sky, but away from the direction of the Sun. Look for signs of false colour on either side of the object. Slight colour fringing, usually apple green and magenta (pinkish), is inevitable, but it should be noticeable only if you look hard for it. Turn slightly so that the object moves to the edge of the field of view. The object should remain sharply in focus.

Cheap binoculars usually have a limited field of view – that is, the image you see is surrounded by a large area of black, like looking into a tunnel, rather than giving you a wide perspective as in the more expensive models. But a wide field of view is useless if the definition is not good all the way across. Avoid wide-field eyepieces that may give a curved plane of focus. Check for this by examining a flat wall at right angles to you – it should be perfectly in focus from edge to edge. Also check that the magnification is even across the field of view by panning across it – avoid binoculars with barrel distortion, which makes straight lines near the edges of the field bow outwards, and will make stars near the edges of the field of view seem to bulge towards you as you sweep across the sky. Check for vignetting (uneven illumination of the field of view) by looking through the binoculars the wrong way. You should see the circular outline of the eyepieces even when looking through from the very edge of the objectives – the outline should not be cut off by the edges of the prisms inside. Very lightweight but cheap binoculars may not be rugged enough to prevent the prisms from becoming misaligned by an accidental knock, and you could be faced with a costly repair bill for realigning them.

If you wear spectacles for short or long sight, make sure you can focus on infinity using the binoculars with your spectacles off. But if your spectacles are to correct astigmatism, you will have to keep them on while observing and you will need binoculars with rubber eyecups which can be turned back to give a soft surface against which you can press your spectacles.

There is also the question of *exit pupil* to be considered. The exit pupil is the diameter of the beam of light that comes out of the eyepiece and into your eye. For a pair of binoculars it is easily calculated – it is the size of the objectives divided by the magnification. So the exit pupil for a pair of 10×50s is 5 mm, and for a pair of 7×50s it is just over 7 mm. You can see the difference if you hold the binoculars away from you so that you can see the disks of light in the eyepieces.

When your eyes are dark adapted, the pupils open up from their daytime diameter of about 3 mm to about 7 mm. This figure is an average for the population at large, and values for individuals vary from about 8.5 down to 4 mm, even among young, healthy people. There is a definite general decline with age, and the average for eighty-year-olds is about 5 mm. There is no point in using 7×50 binoculars, which have an exit pupil of 7 mm, in order to see a brighter image if your pupils only open to 5 mm. So it could be worth you checking your pupil size before you buy. The simplest way is to get someone to measure them with a ruler held up close to your eye in dim light.

Even experienced telescope owners still use their binoculars, so you should expect to get good use out of them. They are fine for locating such objects as comets and novae, and simply for the pleasure of gazing at the Milky Way. Binoculars can reveal a good range of interesting objects, even from the city. But most amateur astronomers find that the low-power views through binoculars are not all they want, and inevitably they begin to think of getting a telescope.

Which telescope?

Astronomers love to talk about telescopes and to compare one with another. And, as in any specialist discussion, there is as much nonsense talked as there is truth. When it comes down to it, the "best" telescope is as impossible to define as the "best" car. In astronomy, as with cars (and other things), it's not what you've got, it's what you do with it that counts.

Of the three basic types of telescope – the refractor, the reflector, and the catadioptric – each has its own peculiarities and advantages. Even so, whatever type you have and wherever you live, you can still accomplish a great deal. But if you want to get the most out of city observing, it helps if you use the most suitable instrument. Because so many people ask for guidance on buying a telescope, I shall go into some detail about the basic types and what they can do.

Refracting telescopes

The refracting telescope is the popular concept of what a telescope should look like, with a big lens – the objective – at one end pointing skyward, and a small lens – the eyepiece – at the bottom of the tube, through which you observe. The objective focuses light from distant objects to form an image, and the eyepiece gives a magnified view of the image. To magnify it more, you use a more powerful eyepiece. The size of a telescope is usually indicated by the diameter of the objective, as the magnification it gives can be changed by using a different eyepiece. A magnification of 100 times, say, is often called a power of 100 or 100 power, and is often abbreviated to ×100, or 100×. The higher the magnification, the higher the power you are using. Figure 5.1 shows a typical modern refractor.

There is one drawback that prevents most refractors from being the ideal telescope: *false colour*. A single lens, such as a magnifying glass, splits light into the colours of the rainbow as it bends or refracts it towards a focus. This is not very noticeable in a magnifying glass, which magnifies only a few times, but in magnifying the image many times, as in a telescope, the false colour becomes objectionable, and some means has to be devised to reduce it. The quality of a refractor is to a large measure determined by how, and how successfully, this is done.

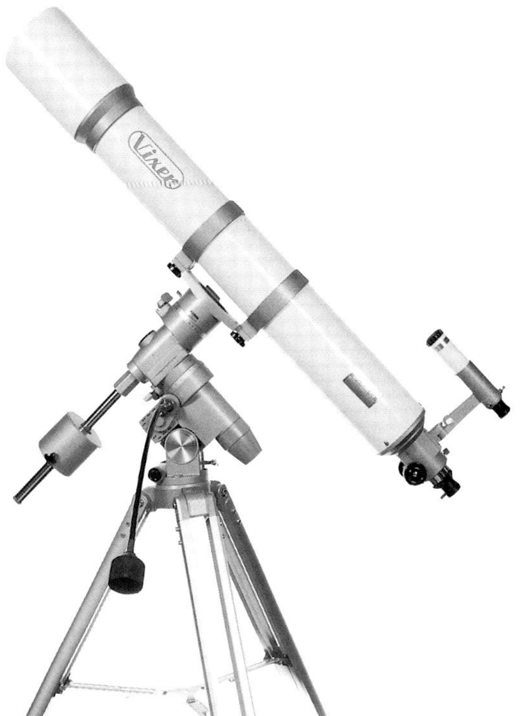

Figure 5.1 *This Vixen 100-mm (4-inch) refractor is on a Great Polaris mount, a precision-built equatorial mount with a built-in pole-finding system.*

Cheap telescopes sold by a wide range of outlets from mail-order houses to department stores are what is called *non-achromatic*: they have only a single-element objective lens, which itself does nothing to correct for the false colour it generates. The aperture is generally *stopped down* – that is, there is a disk with a hole in it some distance down the tube, which considerably restricts the instrument's effective diameter. The purpose of this stop is to reduce the inevitable false colour in the image. But it also makes the image so dim that most of these telescopes fail to show any but the brightest stars. The result – disillusionment with astronomy. I mention such telescopes only because they are widely available and often carry quite a high price-tag: about the same as a reasonably good pair of binoculars or a small portable black-and-white TV. You certainly cannot go by the external finish where telescopes are concerned.

The next stage in improving false colour is to make the objective lens from two glass elements – a *doublet* – arranged so that the colour dispersion of one counteracts that of the other. Such a lens combination is called an *achromatic lens*, or simply an *achromat*. This is the system used in a wide range of refractors costing anything from the price of a portable CD player to the price of a budget car. At the cheap end the quality often leaves something to be desired: either the false colour is not properly corrected, or there are other image imperfections which make the telescope suitable only for very low magnifications, below ×20.

Even well-made basic refractors suffer from some false colour. With the most readily available types of optical-quality glass it is possible to correct for only some of the false colour, and in my experience a typical refractor gives objects a bluish halo. The better the telescope, the less objectionable the halo and the higher the magnifications you can use. A 75-mm (3-inch) refractor of this type will cost about the same as a large colour TV, and should give crisp views up to a magnification of ×150.

The ultimate in colour correction in refractors comes from using exotic materials such as fluorite in place of glass for one of the elements of the objective lens. The more expensive of these telescopes can give virtually colour-free images, and have price tags similar to those of small cars. Some telescopes described as "fluorite" use a conventional two-element lens with an additional fluorite corrector. Objective lenses that give excellent colour correction are called *apochromatic* or simply *apo* lenses.

Reflecting telescopes

In a reflecting telescope the image is provided by a concave *primary mirror* rather than a lens. The great thing about mirrors is that they reflect all colours equally, so there is no false colour in the image. However, the image is formed right in the middle of the incoming beam of light, so a smaller *secondary mirror* is placed so as to reflect the light to the side of the tube, where the image is viewed through an eyepiece. This basic design of reflector is called the *Newtonian* (see Figure 1.3 on page 10).

This arrangement has important consequences. First, you view sideways on to the object you are observing, which does not affect your view in any way, but makes finding objects a little more tricky than with a refractor. In order to aim it you have to shift your observing position to squint along the tube or look through the finder telescope, which is usually a low-power refractor. Second, the flat secondary mirror both blocks a little light from the main mirror and introduces the effects of diffraction into the image. *Diffraction* is a slight bending of light as it passes an obstacle, in this case the secondary mirror and its support. Because a small proportion of the incoming light is diffracted, and does not go where it is wanted, the contrast of the image is reduced a little. It is like turning down the contrast on your TV set.

Third, a little light is lost with each reflection. While this does not matter too much when the mirrors are freshly coated, after a while the percentage of light reflected will decline from about 88 percent when new to, say, 82 percent as the coating's reflectivity deteriorates over a period of a few years. Since there are two mirrors, the amount of light transmitted to the eyepiece will be reduced from 88 percent of 88 percent, which is about 77 percent, to 82 percent of 82 percent, which is about 67 percent. Over time, each mirror will also tend to collect a layer of dust specks, further reducing the contrast and light transmission. Cleaning a mirror is no simple task – the surface coating (usually aluminium) must be treated with care. By contrast, the objective lens of a refractor absorbs only about 2 percent of the incoming light, while reflection from the glass surfaces can be reduced to a similar percentage by the use of antireflection coatings. The total light transmission of a good refractor is in excess of 90 percent, and, unlike that of a reflector, it remains close to that figure. Any dirt on the lens is comparatively easy to clean off.

The tubes of reflectors are usually open at the top end, providing access not only for dust and other airborne particles, but also for stray light. Air currents are far more likely to circulate in the open reflector tube than in the sealed refractor, so a reflector takes longer to settle down when it is carried from indoors out into the night air. And the mirrors of reflectors can easily go out of alignment, making them less straightforward to use and maintain than refractors.

All these drawbacks would seem to make reflectors a poor choice, but they have two saving graces. First, they are much cheaper to make in larger sizes than are refractors. Second, the lack of false colour means that a reflector can give a superior image to many refractors. For many amateurs, therefore, the reflector is the choice. The same amount of money that will buy you a rather average 100-mm (4-inch) refractor will get you a 150-mm (6-inch) reflector that will show finer detail on planets and fainter deep-sky objects. The price differential becomes much greater as you go to larger sizes.

Catadioptric telescopes

Both refractors and reflectors are rather long for their diameter. This feature, broadly speaking, is called the *focal ratio* or *f-ratio*. Properly defined, it is the ratio of the focal length of the objective lens or mirror to its diameter, the resulting number being known as the *f*-number. (The

Figure 5.2 *A Meade LX-200 250-mm (10-inch) Schmidt–Cassegrain. The telescope is controlled by a purpose-built computer which can locate and follow objects, even when used as an altazimuth instrument, as shown here.*

focal length is the distance from the lens or mirror to the focus, the point where the image is formed.) So an *f*/15 telescope is very long compared with its diameter, and an *f*/4 telescope is short and squat. The squatter the telescope, the easier it is to handle and to mount. But there is a price to be paid for this convenience: such telescopes are more expensive, and perform differently from longer *f*-ratio instruments. In general, the shorter the telescope, the lower the magnification it will give with a particular eyepiece. But the catadioptric design, which combines reflecting and refracting optics, overcomes this problem to give a short-focus but powerful telescope that is easy to mount and to carry around.

The most common design of catadioptric is the *Schmidt–Cassegrain telescope*, or SCT, as popularized by manufacturers Celestron and Meade (see Figure 5.2). A telescope with a standard 200-mm (8-inch) mirror would normally need a tube 1.5 to 1.75 metres (say 5 to 6 feet) long. An SCT has a glass corrector plate – a thin lens – at the top end of its tube, closing it in. The secondary mirror reflects incoming light back down the tube (not to the side, as in the Newtonian) and through a hole in the main mirror, behind which is located the eyepiece. You observe in the same direction as the object, as in a refractor, but the tube is only some 600 mm (24 inches) long, even though the telescope is *f*/10.

The SCT design packs a lot of telescope into a small package, and because of its compactness the mounting can be a lot lighter than for a refractor or reflector of similar aperture. A whole industry has grown up around making accessories for these popular telescopes.

An SCT works out at nearly twice the price of a Newtonian reflector of the same aperture, but many amateur astronomers find that the convenience of the design is worth it. Optically, though, there are shortcomings. The worst is that the secondary mirror is larger than it would be in a Newtonian, so the effects of diffraction start to become more evident. Contrast may be worse than in a Newtonian, and the optics are more difficult to realign if they become disturbed. There are many SCTs around that give indifferent images for this reason. Few dealers in the US appear to check the alignment before they ship them to customers. In the SCT's favour, however, is that the closed tube should keep the main mirror free from dust, make the instrument quicker to settle down when taken outside, and reduce air currents in the tube.

Other essentials

A good telescope has many essential components besides its optics, and a shortcoming in any one of them can destroy its usefulness. The experienced observer knows that a telescope's mounting is vital. Many cheap telescopes are let down as much by an insubstantial, wobbly mounting as by other failings. Some amateurs maintain that the eyepiece is just about the most important part of the telescope, and indeed it is true that a good telescope can be ruined by a bad eyepiece. In each case, though, it is rather like claiming that one of the legs of a three-legged stool is more important than the other two.

Mountings

Ideally, what you want from the telescope's mounting is that you should be able to swing the instrument so as to bring it to bear on your target,

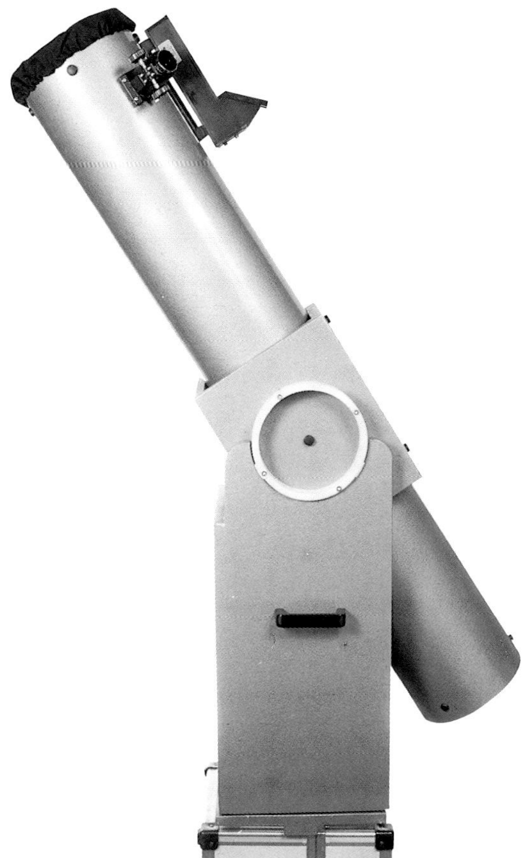

Figure 5.3 *Dobsonian reflectors such as this 200-mm (8-inch) Dark Star instrument offer low-cost observing. The altazimuth mounting is more suited to deep-sky than to planetary work, and photography is limited to short exposures. This telescope is equipped with a Telrad finder, at the top of the tube.*

and find that target within seconds. The drive, a mechanism that slowly turns the telescope to compensate for the apparent rotation of the sky, should be sufficiently accurate for you to follow the target for as long as you wish. You should be able to focus easily without the telescope taking minutes to stop shuddering, and a gentle breeze should be able to waft past without the image breaking up into dancing patterns like a laser light show.

There is always a tendency to keep costs to a minimum, and a telescope that is manufactured to a price cannot be expected to be rock steady. I shall not discuss mountings in any great depth because the choice has little to do with whether you observe from the city or from the country. The main thing is to make sure that the mounting is well enough engineered for your observing not to be restricted by it.

Your basic choice is between the simple *altazimuth* mounting, which allows the telescope to move up and down (in altitude) and from side to side (in azimuth), and the *equatorial mounting*. In this, one axis, the polar axis, is aligned on the celestial pole, so that motion of the telescope

Figure 5.4 *A 20-mm Nagler eyepiece (left) and a conventional 20-mm Erfle eyepiece. The Nagler, with its 50-mm (2-inch) barrel, gives an apparent field of view of 82°, while the Erfle, in a 31.7-mm (1¼-inch) fitting, has an apparent field of view of 60°. A similar-sized 25-mm orthoscopic eyepiece, by comparison, has a field of view of only 36°.*

about the other axis is parallel to the celestial equator, like the apparent movement of the sky, enabling objects to be tracked more easily.

The advantage of the altazimuth mounting is that it is cheap, so you can buy a steadier mounting for your money. From an engineering point of view, the stresses are easily dealt with, and these days the very largest professional telescopes have altazimuth mountings. The popular *Dobsonian* telescope, shown in Figure 5.3, is essentially a Newtonian reflector on a redesigned altazimuth mount which is easy to make. Its main advantage is its cheapness, and it is an excellent way of mounting a large instrument at low cost. As a functioning telescope, however, it is not particularly portable and is best suited to fairly low magnifications. Having said that, I would rather have a good 150-mm (6-inch) Dobsonian with a smooth and steady mount than an average 100-mm (4-inch) refractor on a shivery tripod and equatorial mount, even for planetary observing.

Eyepieces

Tried and tested eyepiece designs include Kellners, orthoscopics, Erfles, and Plössls; for small refractors, Huygenians and Ramsdens are also used. (For a simple outline of the various eyepieces, refer to *Norton's 2000.0*.) A good rule of thumb is that if your telescope is *f*/8 or longer, you can get away with simpler eyepieces than if it is *f*/6 or shorter. A simple and cheap eyepiece design such as the Kellner is quite adequate for much general viewing, particularly at the lower powers and with fairly long *f*-ratios. For the higher powers, orthoscopics are also perfectly suitable. An Erfle eyepiece offers a fairly wide field, though the definition at the edge is not very good. Plössls can give better definition over a wider field than orthoscopics, but the name has sometimes been applied in rather cavalier fashion to inferior designs.

Small telescopes take eyepieces 24.5 mm (0.965 inches) in diameter, while most amateur telescopes in the 150–250 mm (6–10 inch) range use 31.7-mm (1¼-inch) eyepieces. For basic eyepieces less than about 18 mm (¾ inch) in focal length, the smaller barrel is all that is needed, but the 24.5 mm size will not give very wide fields of view.

In recent years, new ultra-wide-field eyepiece designs have come along, notably the Naglers (see Figure 5.4). These can give a field of view of 80° or more, and some observers will use nothing else, even for planetary work where the wide field is not important. Some ultra-wide-field eyepieces have 50-mm (2-inch) barrels, so if you plan to use them you will need a focusing mount of this size, and an adapter for when you want to use a standard eyepiece.

Finders

A finder is a small telescope with a wide field of view, attached to the tube of the main one and used to locate your target. Do not underestimate the value of a good finder, for it is even more important for city observing than in the country. If the skies are dark you can aim your telescope at roughly the right spot fairly easily, because you can see the fainter stars with the naked eye. This is where non-magnifying finders, such as the Telrad, come into their own. As Figure 5.5 shows, they provide an illuminated reticle projected on the sky, so that you can aim the telescope just where you want it. If you cannot see even the faint stars, however, you

Figure 5.5 *The view through a Telrad finder, which projects rings into the line of sight over the telescope tube. This helps you to align the telescope on faint objects whose position you know relative to stars that are visible. The rings appear to be at infinity and can be viewed either by eye or using binoculars, which is particularly useful in city skies.*

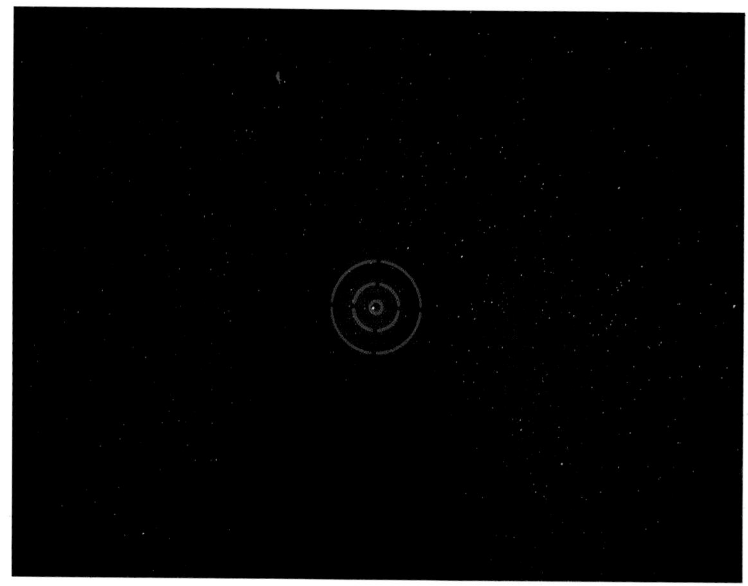

Figure 5.6 *A Super Polaris mounting. To align it, first set the scale just above the eyepiece to your distance east or west of the meridian of your time zone. Then turn the polar axis until the date (months are indicated by 1–12) is next to the time (indicated by 1–24). In the northern hemisphere, turn the mounting so that Polaris is in the field of view of the eyepiece (inset), then centre it in the small circle on the graduated line. In the southern hemisphere, centre Sigma Octantis and the other nearby stars in the other small circles. The mounting will now be aligned.*

will need a low-power finder. A low power and wide field of view can make all the difference when seeking that elusive deep-sky object.

Photography through the telescope

To take photographs through the telescope, you need an equatorial mount whose polar axis is perfectly aligned on the celestial pole (making it parallel to the Earth's axis), a smooth and accurate drive system, and a means of checking for and correcting the drive errors that inevitably creep in. These days you can buy well-engineered mountings that will drive accurately, powered by ordinary batteries, and including a hand-held drive corrector with push buttons for making slight adjustments to the drive rate. To monitor the exposure, you need either a finder or what is called an off-axis guider, which allows you to view a star outside the field of view of the object being photographed.

Aligning the polar axis on the pole is made easy on the many commercial mountings which have pole-finding systems, as shown in Figure 5.6. For the best results, however, it is best to have the mount permanently fixed in place. That way, you do not lose valuable observing time in setting up. Too often I have either finished setting up just as the clouds have rolled over, or taken a chance on a quick alignment that subsequently proved to be slightly out. The longer the focal length of your telescope and the longer the exposures you plan to do, the more accurate your alignment needs to be. For exposures of a few seconds, either on the planets using film or on deep-sky objects using a CCD, rough and ready alignment may be good enough.

Making your own

Amateur astronomers have enjoyed making their own telescopes ever since William Herschel decided he could not afford the telescopes on sale in eighteenth-century Britain, and found he could make a better one himself. It is only recently that the telescope market has expanded to the point where you can expect to go out and buy an instrument that will do all you want at a price you can afford. Even so, there are plenty of amateurs who make their own, either because they cannot afford a ready-made telescope or just because they get a kick out of doing it.

If you want to go all the way you can even grind your own mirror. At one time every self-respecting amateur was expected to have had a shot at grinding a mirror, but these days it is mostly worth doing only for the fun of it, not because it is going to save you money. You start with two bits of glass, called blanks, one of which is to become the mirror, and the other to act as the grinding tool. In a controlled fashion, you rub the mirror over the tool with a grinding grit in between, and eventually the mirror becomes concave and the tool becomes convex. By using finer and finer grinding paste you get a smoother and smoother finish, The final stage is to polish the mirror to the correct paraboloidal figure. To test the figure of the mirror, you make a simple rig from a light bulb and a sharp edge. Amazingly, this set-up allows you to check the figuring of the mirror to an accuracy of a few tens of nanometres.

With smaller mirrors, you will find that buying the materials costs you virtually as much as if you had bought one off the shelf. For the larger sizes, it depends on how much you value your own time. If you know someone who will aluminize the mirror for you cheaply, the equation may well begin to tilt in your favour. You may even find that silver-coating the mirror yourself is feasible, though the technique is rather a forgotten art. Silver is now cheaper than it used to be, so silver-coating at home could prove more economical than commercial aluminizing. The main difficulties are obtaining the right chemicals and preparing the surface of the glass so that it will accept the silver coating. Silver is more reflective than aluminium – 93 percent when fresh, compared with 88 percent. However, it is attacked by sulphurous acid, which is a major component of industrial pollution, and can tarnish within weeks under unsuitable conditions. So any economy you achieve by home coating may be offset by the need to resilver the mirror more frequently.

Having made or bought the optics, you can build a Dobsonian mount in a couple of weeks for the basic unit, working in the evenings. You need the ability to saw, drill, and fasten together components made from plywood and other materials. The more accustomed you are to DIY work, the easier it will be. Making your own equatorial mount depends on what facilities you have available. Unless you have some experience of engineering and access to a reasonable workshop, it is probably best to limit yourself to a Dobsonian. Plenty of people who have "made their own telescope" turn out to have persuaded someone else to do the difficult bits. You can either make a straightforward mount or you can branch out with ideas of your own. Everyone interested in telescope making is full to the brim with original ideas, most of which were actually thought of fifty or more years ago and forgotten. Every so often, though, something really new does come along.

The advantage to the city astronomer of a home-built telescope is that it is possible to incorporate the contrast-enhancing features described later in this chapter, though these may require more work than is usually put into a Dobsonian-style instrument. For details of telescope-making techniques, see the books listed in the Bibliography. New ideas and designs are featured every month in *Sky & Telescope*'s "Telescope Making" column.

The city astronomer's choice

The one type of telescope that is not particularly good for visual observing from towns and suburbs is the very short-focus reflector. Really short *f*-ratios, such as *f*/4, are best suited to observing deep-sky objects at fairly low powers, with the potential for wide fields of view. In a bright sky, much of this potential is wasted. This is not to say that a short-focus reflector is useless – it is just not the most appropriate instrument for the visual city astronomer. However, for photography or CCD work, its additional speed and wide field are a definite advantage.

Some people would advise, with justification, that the best telescope for city use is a perfectly corrected refractor of around 150–180 mm (6–7 inches) aperture. Such a telescope is ideal for observing those objects least affected by light pollution, namely the planets. A good refractor gives the highest possible contrast, which is important if you want to see fine planetary detail and also if you wish to search for small deep-sky objects. The modern designs allow for quite low *f*-numbers (around *f*/7 or *f*/8), which means that you can use fairly wide-field eyepieces. The shorter tubes of these telescopes makes them not so massive as to require a permanent observatory.

But there are usually other considerations, notably cost and space. For most people, these matter a great deal. Not only is a top-quality 150-mm (6-inch) refractor expensive, but it needs to be mounted quite high so that you can view through the eyepiece without lying on the ground. The pedestal will probably need to be as high as you are. Unless you are either dedicated or fortunate enough to own one, you will have to choose between less expensive and possibly more portable instruments. What about standard Newtonian reflectors?

I shall nail my colours to the mast here and say that my own preference is for the reflector. While refractors are traditionally noted for their better contrast and image stability, I have always thought that a reflector is a much better choice (unless money is no object). You get more aperture for your money with a reflector, and it is usually easier to observe with. As for the supposed lack of performance, in my experience what counts is focal ratio. The higher the focal ratio, the smaller the secondary mirror compared with the main mirror, and the smaller the central obstruction. There will also be fewer traces of coma, which is the tendency for star images towards the edge of the field of view to appear as small streaks instead of points of light.

In practice, a well-made reflector will deliver superb planetary images at a fraction of the cost of a refractor of similar aperture. Although, size for size, the refractor has the edge, the difference is much less than it was once held to be. In a detailed test report published in *Sky & Telescope* (March 1992, pages 253–7), Douglas George concluded that a good-quality 150-mm (6-inch) reflector approached the performance of a 175-mm (7-inch) refractor. And since the reflector costs much less, it wins my vote. The key is in the quality of both optics and manufacture. Do not buy the cheapest reflector available and expect top performance. If you are prepared to spend more, you should get excellent performance at a fraction of the cost of a fluorite refractor. If you have the choice, aim for an *f*/8 instrument rather than *f*/5 or *f*/6.

There is a further option. The main advantage of reflectors over refractors is that they can be made in larger sizes without the cost rising steeply. You can buy commercially made Dobsonians of enormous aperture for comparatively little money: telescopes of 430 and 500 mm (17 and 20 inches) aperture are readily available at about twice the price of a basic SCT. The optics may not be perfect, but the aim is to bring in as much light from deep-sky objects as possible – the so-called "light bucket" approach. While this may seem to be pointless in a city environment, there are advantages. After all, if the objects you are searching for are faint anyway, the more light you can pull in, the better. So the very large reflector also has its place in the city astronomer's armoury.

David Cortner, who observes from a light-polluted part of Tennessee, owns both a 400-mm (16-inch) *f*/5 reflector and a 125-mm (5-inch) *f*/6 Astro-Physics fluorite refractor. He says:

> As a card-carrying member of the "A-P [refractor manufacturer Astro-Physics] League," let me say that the best views I get from my light-polluted backyard come from my light bucket, not from my refractor. The only way you'll see galaxies looking like galaxies is with aperture, and aperture is not as crippled by suburban light pollution as I expected it to be. My 16-inch will get 15 to 16th mag. easily in the country, but from the city it still pulls in 14.0 to 14.5, even when the sky is pale and unexciting. Vaguely quantitatively: the naked-eye light-pollution tax here is maybe three magnitudes, but the telescopic penalty is nowhere near as severe, and is less so the more aperture I bring to bear.

What about the SCT? There is no doubt that the contrast is not as good as in a refractor or long-focus reflector, despite the many testimonials to the contrary that the manufacturers use in their advertising. However, the real advantage of the SCT lies in its convenience and the availability of accessories. You will get better performance with a well-made and well-maintained SCT than with a reflector with a dirty mirror. Computer-controlled SCTs are now becoming very popular. With very little initial alignment, you simply centre a known star in the field of view, repeat with another across the sky, and then request the computer handset to locate any object in its vast memory bank.

A telescope for the city

Whatever telescope you own, there are ways of optimizing your telescope for city life, in particular by making sure that it gives maximum contrast. The name of the game is *signal-to-noise ratio*. The signal is the feeble light from the object, and the noise is the vast flood of light from the sky in general, from streetlights, from security lights, and from your neighbours' homes. Your task is to maximize the signal and minimize the noise.

Keep it clean

Getting the maximum signal means keeping your optics as free from dust as you can. Use a tight-fitting cover that keeps all dust out when the telescope is not in use, and line the cover with blotting paper or some other absorbent material that will soak up any dew that may be on the optics after a night's observing. There will need to be an air hole in the cover so that it can dry out, or you will be trapping the moisture. A good material for making sure that the cover fits tightly is self-adhesive velour, used for the bases of table lamps, and available in rolls in hardware stores.

Inevitably, dust will collect on the surface of mirrors and lenses. And it doesn't just lie there – it gets stuck on. Do not try to blow it off, for you will just end up spitting on the glass. You can try a can of compressed air, as used by photographers to clean negatives, but these can leave residues, and often they fail to remove the dust anyway.

A little dust can be lived with. If you are happy with the results you are getting, it may be better to leave the dust alone than to try cleaning it off. Eventually, however, the day arrives when it has to go. If you have a refractor or an SCT, use a proprietary lens-cleaning fluid and a clean handkerchief to remove it. There should never be any need to remove the objective lens from a refractor, so that is all the maintenance to the optics you will need to do. A mirror, however, has a more delicate coating, and you will have to remove it from its cell, which usually means removing the cell from the telescope first. You should clean your mirror at least once a year if there is an obvious build-up of dust. Cleaning actually prolongs the life of the surface. Dust absorbs moisture and acts as a site where contaminants can gather and eventually corrode the mirror. This is why tiny pinholes eventually appear on a mirror, even if you never touch it.

Immerse the mirror in warm water in which a small amount of wetting agent has been thoroughly dissolved. Photographer's wetting agent is ideal; you can use a non-biological detergent such as Dreft. Do not use dishwashing liquid as this often contains other ingredients, such as lanolin, which make it gentle on the hands but can leave a greasy residue that may well be worse than the dust you had in the first place. In hard-water areas, use distilled or deionized water as supplied for car batteries or steam irons.

Even after a good soaking the dust may be reluctant to shift, so you will have to swab the glass gently with fresh cotton wool, changing it frequently to make sure that you do not drag sharp particles across the surface. Finally, choose a dust-free place for the mirror to dry. Support it resting on its edge on a wad of clean absorbent paper, so that the water drains off quickly and is soaked up.

Now you will have to realign the optics. You have to make sure that when you look through the eyepiece hole, the outlines of both the flat and the main mirror are concentric, and that the reflection of the flat in the main mirror is dead central. It is a good idea to put a small spot at the centre of the main mirror to help you align the optics. It will always be in the shadow of the secondary, so there is no need to remove it afterwards. Alternatively, make a cross of string at the top and bottom of the tube to help you get everything in line. Realignment is a detailed procedure, and is dealt with more fully in books listed in the Bibliography.

Down with noise

Assuming that your optics are clean, you now need to cut out unwanted stray light – the "noise." For all telescopes, a good long *dew cap* or tube extension is essential for city observing. With refractors and SCTs, it shelters the objective or corrector plate and helps prevent the formation of dew, and with both refractors and reflectors it helps to prevent stray light from entering the tube. This is particularly important with Newtonian reflectors, as you will see if you look into the eyepiece tube at night with no eyepiece present. As well as the rather small area occupied by the

Figure 5.7 *Ways of improving the contrast and performance of a reflecting telescope. Extending the tube and painting the inside of the tube black are the simplest and most effective measures. Baffles cut down the amount of stray light that reaches the main mirror. The fan and the optical window are alternatives: a fan is not needed if the tube is closed by a window, with the tube also blocked at the lower end. A fan installed behind the main mirror draws through a steady flow of air, preventing tube currents from building up and disturbing the seeing.*

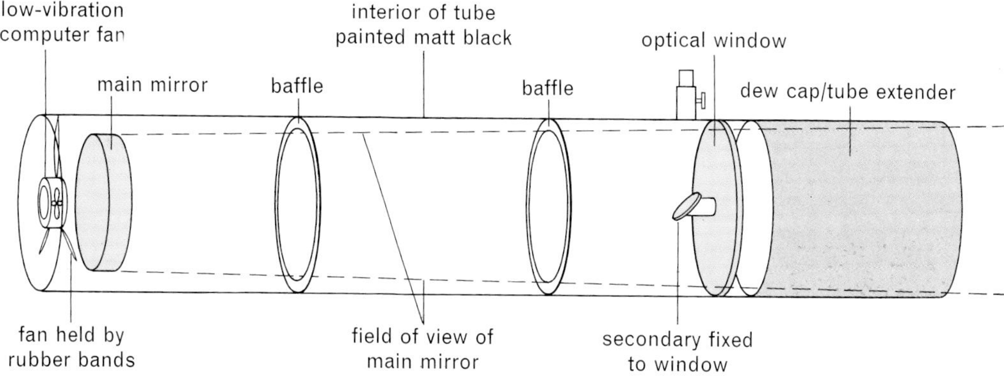

reflected image of the main mirror, you can see a large expanse of the opposite side of the telescope tube, the support for the secondary mirror, and maybe some of the focuser itself. Around the reflection of the main mirror there may be visible part of its cell. Ideally you should not be able to see these parts of the telescope at all. They are just contributing stray light which reduces the contrast of your image.

By fitting a dew cap or tube extender, you are restricting the light that enters the telescope more closely to that from the area around the object itself. Make sure the dew cap does not encroach on the field of view of the telescope. If it were the same diameter as the telescope tube and were too long, it would reduce the effective aperture. A dew cap and other ways of fine-tuning your telescope's performance are shown schematically in Figure 5.7.

Inevitably some stray light does get in, and steps should be taken to prevent this light from reaching the mirror or eyepiece. The inside of the tube can be painted matt black, but make sure that the paint really is matt. A smooth surface such as PVC is really quite difficult to dull down, and the slightest shininess in the paint will create reflections. If you can manage it, it is a good idea to fit internal ring-shaped baffles to cut down any further reflections. These are often fitted to refractors, but less often to reflectors. Do not, however, use the self-adhesive velour mentioned above to line the tube. This will provide an insulated layer around the inside of the tube which will retain heat, giving you trouble from tube currents.

Alan MacRobert has used a novel method of blackening the inside of his 320-mm (12½-inch) reflector tube (*Sky & Telescope*, December 1992, page 696). He painted the inside with glue, on which he scattered handfuls of sawdust. When it was dry, he sprayed the whole thing with matt black paint, taking four cans to complete the job. He suggests that a more efficient method would be to dab the paint on with a sponge on a stick. As a result, the sawdusted area was truly black, though the lower parts of the tube, painted only, were still shiny. If you have an open or skeleton tube to your telescope, enclosing the tube in black plastic sheeting will be a good start, though this is reflective and not the ideal material.

Rich Combs, who observes from San Francisco's Bay Area, advises that to approach refractor performance with a reflector it is also necessary to attach a flat window at the top of the tube. Ordinary flat glass rarely has flat and parallel sides, so this glass needs to be optically worked to eliminate distortions. A window on a reflector is advantageous for several reasons: first, it will cut down the currents that swirl around inside the tube when the telescope is not at ambient temperature; second, it will prevent dust from getting onto the main mirror; third, if it is located just above the eyepiece (and well inside a dew cap or tube extension) it will allow you to do away with the vanes of the assembly which carries the secondary mirror. You can fix the secondary assembly directly to the inside of the window, and then adjust the angle of the secondary by slight movements of the window. Rich sums up:

> In general, I would say for visual use, about 80 percent of the improvement you will notice in the scope's performance will be gotten from the first 20 percent of the effort you put into limiting stray light. This

involves things like extending the input end of the tube, lining the interior of the tube with a photon-sticky substance, and closing the mirror end of the tube. The extra effort of fully baffling the scope will get you most of the other 20 percent.

Designs for better contrast

One source of low contrast in reflectors is the size of the central obstruction caused by the secondary mirror. Not only does it obstruct some light, it also degrades the image somewhat through diffraction effects. As shown in Figure 5.8, the image of a star consists of a tiny false disk, surrounded by spikes (and, in larger instruments, rings) produced by the diffraction of light as it passes through the circular aperture of the telescope, whether it be a refractor or a reflector. The effect of the central obstruction of the reflector is to throw slightly more light into the spikes and slightly less into the disk. If you imagine a planet as a multitude of point sources of light, then a perfect image of it would reproduce each of those points perfectly. But if the image of each point is a tiny disk surrounded by diffraction spikes, this will cause a very slight smearing of the detail. However, Horace Dall, whose knowledge of telescopes was unrivalled, maintained that as long as the diameter of the central obstruction was less than 20 percent of that of the main mirror, the resultant loss of contrast was negligible. Nevertheless, the conventional SCT has a central obstruction of this order (see Figure 5.9), and the loss of contrast compared with an *f*/8 Newtonian is noticeable.

It is possible to reduce the size of the secondary in Newtonians at the expense of field of view, in order to cut the contrast loss to a minimum. If you are a planetary observer and have a good, driven mounting, a wide field of view is unnecessary. Deep-sky observers could do this too, for

Figure 5.8 *The bright star Capella photographed through a 150-mm (6-inch) f/4 Newtonian. The spikes on the image are a result of diffraction caused by the vanes of the spider holding the secondary mirror.*

Figure 5.9 *A Schmidt–Cassegrain telescope has a large central obstruction – in this Celestron C8, it spans more than 30 percent of the aperture. This reduces the contrast of the instrument's images compared with those from a standard Newtonian of similar aperture.*

many deep-sky objects are small, but if the field of view is reduced to a minimum it will be very hard to locate the objects in the first place.

The vanes, or "spider," which support the secondary mirror also cause diffraction effects which rob the image of some contrast. These are the source of the cross-shape in the images of bright stars, particularly in photographs. One solution would be to make the vanes very thin, but this can make the secondary subject to vibration. Another approach is to use elaborately curved vanes so as to spread the effect over an area instead of concentrating it into a cross.

William Herschel solved the problem another way, though in his case the loss of performance was caused by the poor reflectivity of the speculum metal (an alloy of tin and copper) he used for his mirrors. With two mirrors the light loss was considerable, so he observed by simply tilting the main mirror and looking down the tube. This method was suitable only for his large telescopes; the small ones, such as the 7-foot (2.1-metre) long instrument I was once given the chance to observe with, were of the conventional Newtonian design.

With a large enough telescope, however, you can imitate the Herschelian system. A 300- or 350-mm (12- or 14-inch) reflector will have an appreciable gap between the vanes of its spider. Make a mask for the top end so that you look through only one quadrant, and you will have an unobstructed aperture of just under half the original mirror size. Since the seeing is often better with a smaller aperture, you may paradoxically be rewarded with sharper images of the planets than with the full aperture.

Devotees of unobstructed-aperture reflectors use what are known as *tri-Schiefspieglers*. These have three mirrors, two spherical and one flat, which result in a totally unobstructed yet colour-free image, with long focal length. Such instruments tend to be made to special order, but they are renowned for their contrasty images.

In the quest for contrast the eyepiece should not be overlooked. The trend today is towards wide-field eyepieces made of many glass elements which are usually coated to improve their light transmission and contrast. There are older and simpler designs which give particularly high contrast, such as the monocentric, but these have the drawback of giving an unfashionably narrow field of view. They are much sought after by planetary observers. Nevertheless, the wide-field eyepieces are noted for their exceptional performance on the planets, particularly at short focal lengths. Lanthanum eyepieces, recent arrivals on the market, contain special glass to produce very good *eye relief*, even at short focal lengths. Eye relief is the distance from the eyepiece your eye must be for you to see the full image. With short focal lengths this can be notoriously small, while an eyepiece with large eye relief can be used even by wearers of spectacles.

Whatever eyepieces you use, you will lose performance if they get dirty. I am as guilty as the next observer of leaving eyepieces lying around during an observing session to collect dew and then dirt. They should be regularly cleaned using proprietary lens cleaner and lens tissues.

To help you evaluate the effects of these improvements to your system, there is a sequence of stars of known magnitudes near the north celestial pole (see page 167). This sequence will also allow you to monitor changes in sky transparency from night to night.

Keeping your optics clean and baffling your telescope can have the same effect as treating yourself to a little more aperture – so for city astronomers, time spent during the day can be amply rewarded at night.

CHAPTER 6

Choose your ammunition

As the twenty-first century rolls in, even amateur deep-sky observing is becoming increasingly technical with the advent of interference filters and CCDs. But this is nothing new. Since the mid-nineteenth century, observers have been using a technical fix to reveal fainter objects than they could see by eye alone – photography. Even though it is now being complemented by electronic systems, photography will continue to play an important part in the amateur astronomer's armoury.

For everyday purposes, photography still has a considerable edge over electronic imaging. Even though video cameras are now lightweight, comparatively cheap, and convenient, they cannot compete with a throwaway preloaded camera sold in a blister pack when it comes to taking holiday snaps. Each photograph taken with such simple equipment carries tens of megabytes of information, which would take a considerable amount of computer memory to store, not to mention the hardware needed to display the image. For sheer information-carrying power, a tiny rectangle of film has computer storage well and truly beaten. Add to this the cheapness and convenience, and it is clear that photography is going to be around for some considerable time.

In astronomy, the traditional advantage of photography over observing at the eyepiece is that while the eye records only the light falling on it at any one instant, the photographic emulsion coated on the film can, within limits, record light for as long as you continue the exposure, building up an image. This is what makes it possible to photograph objects which are far too faint to be seen by eye. There is also the advantage that the image is recorded permanently, and can then be measured.

But photography need not be used solely to make a scientific record. Even with the simplest of cameras, from your city or suburban location you can take attractive and interesting astronomical photographs. There are some excellent books on *astrophotography*, as it is called, so here I shall just outline the sort of things you can do. If you want details of how to attach your camera to a telescope or what sort of mount to use, look in one of those other books. In practice, trial and error is usually the best guide, but I suggest exposure details that you can start with.

Cameras

At one time it could be confidently said that any good camera is fine for astrophotography, but that is no longer true. Modern cameras are highly automated and leave little to chance – assuming that you are taking the average sort of picture, that is. Astrophotography is quite out of the ordinary, and even some expensive and otherwise excellent cameras are pretty useless when it comes to the night sky. In order for the light from a celestial object to build up an image on the film, the exposure time needs to be several seconds at least. Some cameras do not permit exposures of more than a fraction of a second, on the basis that you cannot hold a camera perfectly still for that long. They may automatically fire the flash to illuminate the view, which is pointless when the object you are photographing is any more than a few feet away. Only cameras with meters sensitive enough to measure low light levels will allow exposures of between 5 and 30 seconds, which is what you need for even elementary astrophotography. One way to discover what your camera does is simply to try it out. You can usually hear the exposure beginning and ending.

You may find it a good idea to use an old camera for astrophotography. Ideally it should have a lens of *f*/2.8 or faster, a B setting for time exposures, on which the shutter stays open for as long as the shutter release is kept pressed down, and a mechanically operated shutter rather than an electronic one, which drains the batteries over the course of the exposure. It does not matter if the light meter uses a battery, however – that usually draws only a low current and may be left on for some time. A single-lens reflex (SLR) camera is more versatile, and is essential for some work.

Figure 6.1 *A close conjunction of Jupiter and Mars in December 1986, photographed in close proximity to the streetlight that illuminates the author's observing site. In winter the willow tree does little to block its glare. The orange object in the sky is a ghost image of the streetlight. Clouds covered the sky seconds after this picture was taken, further emphasizing the problems of amateur astronomy from a suburban location.*

Twilight and constellation shots

The easiest pictures to take, and often the most aesthetically appealing, are simple shots of astronomical objects in the twilight. A crescent Moon, particularly with Venus or another planet nearby, can often be photographed with an exposure time of a second or so, even on ordinary ISO 100 film. You must keep the camera rock steady. Resting it firmly on a wall is good enough for a one-second exposure, but for longer than this a tripod and cable release are more or less essential. Far from avoiding foreground objects such as buildings or trees, make full use of them: an attractive foreground will enhance the picture's interest. An example is shown in Figure 6.1.

The same applies to photographs of the setting Sun, which can be taken without the need for a solar filter when the Sun is low enough for you to look at it directly in safety. Photographs of the Sun setting between buildings or on a distant horizon with objects silhouetted against it are often well worth taking and turn your city location to your advantage. Getting the exposure right, however, calls for some care. Exposure meters are generally programmed to average the bright and dark parts of a scene, often with a weighting which assumes that the centre of the frame is more important than the fringes. However, the very centre of the view may be unmetered, with a blind spot where the focusing aids, such as microprisms, give a clearer view of the image. You can discover if this is the case by checking the exposure readings when a bright light is visible in different parts of the viewfinder (assuming that your camera tells you what exposure it plans to give). If it does, you must bear it in mind when choosing the exposure.

If the Sun is in this blind spot, its brightness will be ignored and you will get a well-metered shot of the surrounding sky and buildings. The Sun may well be overexposed, and if the buildings occupy a large part of the frame they may come out a boring grey against an overexposed sky. A similar shot but with the Sun away from this blind spot will take its brightness more into account, and will be less exposed overall. However, if you want to show detail on the Sun's image, such as a naked-eye sunspot, then even this average exposure will be too great unless its image almost fills the field of view. You must underexpose the picture by at least one stop compared with what the meter reading suggests in order to show any sunspots. The same applies to attempts to photograph the green flash.

Later in the twilight you can photograph the first stars beginning to appear. Wait until you can just begin to see them with the naked eye, and make an exposure of not more than 10 seconds using the lens at full aperture and a slow film, ISO 50–100. I find Kodachrome 64 excellent for this. You have to choose your moment carefully: too early and the bright sky will wash out the picture; too late and you will get only a black sky with tiny pinpricks of the brighter stars. The aim is to get a good deep blue background so that the sky itself is attractive and any foreground objects show up as well. Figure 6.2 compares results obtained with slow and fast film on the same night. With slow film and the right exposure at the right time, even bad light pollution will not begin to show. But use faster film even five minutes after the optimum time, and the result will be disappointing.

You can even include streetlights in your view, as long as they are not too bright. If they are, you will probably get ghost images as a result of

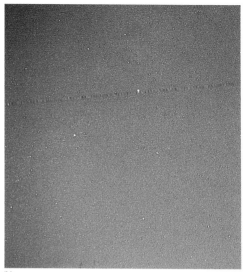
(a)

(b)

Figure 6.2 *Slow or fast film? (a) A 15-minute exposure of Cygnus on Kodachrome 64 compared with (b) a 45-second exposure on ISO 1600 Fuji film. The two exposures are roughly equivalent, but the background is darker and the image more detailed on the slow film. The advantage of fast film is that it permits shorter exposures, placing fewer demands on the guiding system and allowing less time for the lens to dew up.*

reflections within the camera. The same problem can arise if you include the Moon when it is anything larger than a thin crescent. Reflections inside the lens produce a reversed and fainter image diagonally opposite the Moon in the frame. This is more of a problem on fast film. The only solution is to make sure that the Moon is exactly centred in the frame, which may not be very aesthetically pleasing.

Even with slow film, you may find that stars fainter than you could see with the naked eye show up. You may be tempted to give a longer exposure in order to bring out more of them. Try it and see. What will happen, assuming that you are using a standard lens (about 50 mm on a 35-mm camera), is that the stars begin to trail. It is always worth remembering the 10-second rule when taking photographs with a fixed camera. If you double the focal length, you need to halve the exposure time to get the same pointlike star images on the film. Use a wide-angle lens and you can give a longer exposure for the same result. So, by using a 24-mm lens, you can give just over double the standard 10-second exposure time and still get point images. Actually, if you look at them carefully you will see that they are really short trails anyway, but they hardly show as such.

In order to record fainter stars with a fixed camera you must use faster film. Today's fast films of ISO 1600 or 3200 will reveal stars below naked-eye visibility with exposures of 10 seconds at *f*/2 in a dark sky. But from the city the results will be less impressive because of the light pollution. There will be a noticeably coloured or dense background, particularly near the horizon, as in Figure 6.2(b). Nevertheless, if you wait for those really clear nights and concentrate on the area around the zenith, you will avoid the worst effects of light pollution. Oddly enough, it can be more rewarding to take pictures of a brightly moonlit sky. Moonlight produces

Figure 6.3 *Moonlight need not halt your constellation photography. This shot of Perseus was taken with a gibbous Moon in the sky. Moonlight turns the sky blue, just as sunlight does, though the intensity is too low for the eye to see the colour well. On film, however, it tends to overwhelm the colour of the light pollution. Exposure times should be kept to under a minute on fast film, longer on slow film.*

a delicate blue on fast colour film, and tends to overpower the light pollution. Particularly faint stars do not show up, but the results, as in Figure 6.3, can be pretty. Even thin streamers of cloud can lend a romantic look to the picture.

Pictures taken in this way are quite useful for depicting the constellations. Because only the brighter stars are recorded on the film, the main patterns are easier to see. There is an advantage in using 35-mm film, too. Fast film and a standard lens produce star images which are large compared with the distance between them. The constellations show up less well on a larger film format where the star images are more spread out. Some astrophotographers deliberately use a slightly misty filter to spread out the star images slightly, which both brings out the stars' colours and makes the constellation patterns more recognizable.

Driving the camera

Having exhausted the repertoire of pictures you can take with a fixed camera, your next step is to drive the camera so that it tracks the stars. This overcomes the problem of star trailing, and allows longer exposures to be made. You will need an equatorial mount of some kind – that is, a mount with one axis fixed parallel to the Earth's axis. You can buy or make a motor-driven *camera platform*, a mount that will drive the camera at the correct rate, or you can mount the camera piggyback on a telescope which itself is on an equatorial mount. It is also possible to make a simple screw-driven mounting known as a *Scotch mount* or *barn-door mount* (see Figure 6.4). This consists of a pair of hinged boards or plates, mounted so that the line of the hinge points towards the celestial pole. A screw or bolt threaded through one board, which is fixed, pushes against the other board and moves it slowly as the screw is turned.

The trick is to make the device in such a way that turning the screw at the rate of one revolution per minute – in synchronism with the second hand of a watch – drives the movable board at the rate of one revolution per 24 hours. In practice, of course, a full revolution is physically impossible, and after 10 minutes or so of driving the rate becomes inaccurate, but that is about the limit of endurance for hand driving anyway.

The Scotch mount offers a very cheap means of taking driven exposures, but for city astronomers its usefulness is limited to fairly short exposures with standard lenses. The errors in the drive will become obvious if you try to use longer focal lengths. Following the 10-second rule, with a 50-mm standard lens you should aim to turn the screw so as to keep it within 10 seconds of the position of the second hand. With a 135-mm lens, the accuracy should be within 3 or 4 seconds of the second hand, but with a wide-angle lens you can afford to be less precise. Using ISO 400 film and a lens of any focal length at *f*/2.8, exposures longer than 2 or 3 minutes will start to be seriously affected by light pollution. With slower film or lenses, exposure times can be longer. An exposure of 10 minutes on ISO 64 film with a 200-mm *f*/4 lens would require extreme care, but could give good results even in city conditions. Figure 6.5 shows a photograph of Auriga taken using a Scotch mount. The exposure was about 3 minutes on ISO 200, with a 50-mm lens.

However you drive your camera, you will soon find that from the city the limits to exposure are soon reached if you are using just a standard lens. Even a slow film will start to pick up the sky background after a few minutes. Slow film has the advantage that, although you need to expose for longer to record the same image, the contrast of the emulsion is generally higher than for fast film. Images of stars and nebulae are therefore easier to pick out against the sky fog. (*Sky fog* is the term used in astrophotography for the background sky brightness – it has nothing to do with meteorological fog. There is also *chemical fog*, which is the

Figure 6.4 *The Scotch mount is a simple means of driving a camera to follow the stars. This version is designed for northern hemisphere use – the hinge would be at the other end of the boards for use in the southern hemisphere. A simple sighting device parallel to the hinge will enable the mount to be aligned on the pole star with an accuracy adequate for exposures of up to 8 minutes with a 50-mm lens.*

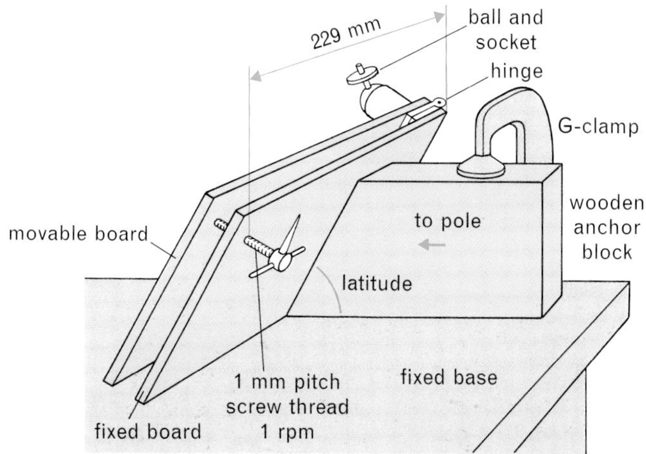

Figure 6.5 *A Scotch mount was used for this picture of Auriga. The method is particularly suited to short exposures, and with film of ISO 200 or faster stars much fainter than those visible with the naked eye can be recorded even from city skies.*

background density you get even in unexposed areas of the film.) You will have to experiment for yourself to discover the best exposure time and film on a given night from your location. It is surprising how many stars you can record in this way, even from a light-polluted location. Objects such as the Orion Nebula and the Double Cluster in Perseus show up well, and even the Milky Way can begin to appear at the right time of year.

So far I have stuck to talking about standard lenses. But, as with visual observing, increasing the magnification for a given aperture darkens the sky background more than it does star images. You could simply stop down your lens from, say, *f*/2 to *f*/4, but it is more appropriate to use a lens of longer focal length, one with (usually) a larger *f*-number than the standard lens.

Bear in mind that to record a faint star you need more aperture, but to record a faint extended object, such as a nebula (or the sky background), you need a smaller *f*-number. Most people do not appreciate that for a given exposure time and film you can record just as much of the extent of, say, the Orion Nebula using a standard lens at *f*/3.5 as you could using the Anglo-Australian Telescope at *f*/3.5. The difference is that the big telescope, with its 3.9-metre (150-inch) mirror and greater focal length, will have a much larger image scale, with the nebula covering about 150 mm of film rather than 0.5 mm or so, and will therefore show much finer detail. And, of course, from the city the fainter extensions of the nebula will be masked by the light pollution.

With lenses of longer focal length, the driving requirements for your mounting and its alignment are more demanding, but with, say, a 200-mm *f*/4 lens and ISO 400 film you can achieve excellent results with exposures up to 10 minutes or even longer. The aim is to find an exposure time that will produce images with a just-acceptable amount of sky background. This will vary widely from night to night, depending on the clearness of the sky.

With a 200-mm *f*/4 lens and ISO 400 film you can photograph any object bright enough to register its image in 10 minutes or so. This will probably include a wide range of open clusters, the brighter nebulae, and some galaxies. You should be able to take attractive pictures of the Orion Nebula, for example, which may encourage you to aim for other objects. However, the Orion Nebula is so much brighter than any other nebula that few others will show up anything like as well. Its only real photographic rival is the Eta Carinae Nebula (NGC 3372) in the southern hemisphere. The Andromeda Galaxy should photograph as an elliptical glow, but the fainter spiral arms may not be revealed. Fainter galaxies such as the Whirlpool Galaxy (M51) will not be detectable. The image scale of a 200-mm lens is too small for it to reveal much of globular clusters, which will appear as small hazy blurs. Even Omega Centauri, the largest and brightest globular, is not very large when photographed with the above combination of lens and film.

The same applies to most Solar System bodies – even the Moon is only 2 mm across when photographed with a 200-mm lens – so I shall deal with the Sun, Moon, and planets separately since they really need to be photographed through a telescope.

Photography through the telescope

Using a telescope for photography is nothing more than using a very long telephoto lens. If you put your SLR camera, minus lens, in place of the eyepiece of one of the popular 200-mm (8-inch) Schmidt–Cassegrain telescopes (SCTs), for example, you are effectively creating a 2000-mm *f*/10 telephoto lens. By the above rule of thumb, this set-up will give an image of the Sun or Moon about 20 mm across (actually about 18 mm/¾ inch) which nicely fits into a 35-mm frame (24×36 mm).

The Sun

Solar photography suffers from the same constraints as solar observing. Just because you are looking through the viewfinder of a camera, do not imagine that the Sun's intensity will not harm your eyes. You must adopt the same precautions as for visual observing. And, as with visual work, the safest procedure is to photograph the projected image.

This poses some problems. A solar projection screen usually gives an image of the Sun some 150 mm (6 inches) across. To get an image this size to fill the frame of your camera you need to be less than about 600 mm (24 inches) from the image. Many standard lenses will not focus this close, and at their closest focusing distance give a rather small image on the film. And even if you can get close enough, the Sun's image will be elliptical rather than circular because you are not at right angles to the screen. The solution is to use a telephoto lens on your camera, with extension rings if necessary to allow you to focus close enough. Having achieved this, you will probably find that the excellent, contrasty projected image which you photograph, with its clear, dark sunspots, turns on film into a pale, lacklustre disk with fuzzy grey spots.

The answer is to use a slow, contrasty film which will emphasize the spots, and to darken the surroundings as much as possible. As with visual observing, an image projected into a darkened shed is ideal (see Chapter 4). This also gives you, on colour film, the most natural colours – which is to say almost none.

The full-aperture solar filters used for visual work are also suitable for photography, and there are glass filters specifically designed for solar photography. The blue image given by the basic aluminized mylar filter is no hindrance when using black-and-white film, but on colour film it gives an aesthetically unpleasing result. An additional yellow filter will cancel out the blue hue, and some commercial filters incorporate one, but beware of overdoing the yellow. The Sun is actually pure white in colour, with a slightly yellow limb. It would be odd if our eyes saw it as anything different, since they evolved to match its output. If the Sun appeared to us as yellow, we would never see a white object by day as white, for it would be the colour of sunlight – yellow. The reason we think of the Sun as yellow is simply that when it is dim enough to be viewed directly, it is low in the sky and is seen through a considerable layer of the Earth's atmosphere. This selectively absorbs the Sun's blue light, leaving it yellow or red.

The Moon

The Moon is a suitable subject with which to begin your photography of the planets. Its size and brightness make it by far the easiest Solar System body to photograph, but it also reveals the problems of planetary photography. Only when you can take good, sharp lunar images is it worth you going on to the planets, which after all are only the same apparent size as individual lunar craters. Anybody can get a passable picture of the Moon, but I have seen some pretty awful pictures of it taken with apertures of 250 mm (10 inches), and some good ones taken with telescopes as small as 75 mm (3 inches).

As for exposure time, it is often said that the Moon should be treated exactly the same as any other sunlit landscape. It does not matter that it is a quarter of a million miles away, just as it does not matter that some parts of a landscape on Earth are a quarter of a mile away – they look just as bright as nearby objects. In fact, if you think of those occasions when you can see the Moon in the afternoon sky, it is obvious that an ordinary daytime exposure will suffice.

Although the Moon is a sunlit landscape, its surface rocks are largely dark basalt, so it does need a little more exposure than a view of the local park. But in general, if you can take a picture on ISO 100 film with an exposure of 1/125 second at *f*/11 on a sunny day, then the same is true for the full Moon. So with an *f*/10 SCT, a good lunar shot would need an exposure of 1/125 second. This is true only when the Moon is high in the sky, so allow a longer exposure if it is lower or if the sky is not perfectly transparent. With an *f*/6 telescope, give about a third as much (that is, six squared divided by ten squared to get the proportion of the areas), or an exposure of 1/500 second. The first-quarter or crescent Moon will need considerably more exposure because the Sun is no longer directly behind you and the landscapes are partly in shadow. Figure 6.6 shows a good example of successful amateur lunar photography.

With exposure times of a fraction of a second, the telescope does not even need to be driven. You should be able to take some perfectly satisfactory pictures of the full Moon in this way. Even if you do not attach your camera to the telescope in place of the eyepiece, but simply point it into the telescope through a low-power eyepiece, you should be able to get good results.

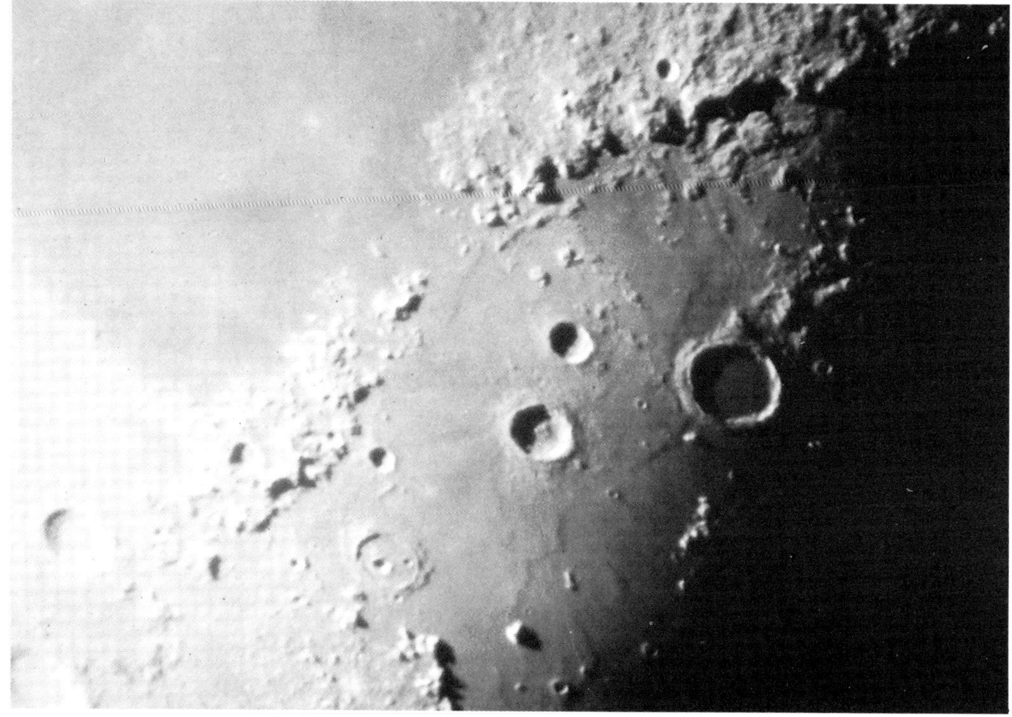

Figure 6.6 *Archimedes (the large crater partly in shadow) and the Apennine mountains, on the fringes of the Moon's Mare Imbrium. The Moon's terminator region is poorly illuminated and needs more exposure than the limb, making it a challenge to photograph, but the results can be spectacular. This picture was taken by Paul Stephens using a 300-mm reflector.*

You *should*, but the pitfalls of astrophotography may now start to become obvious in the shape of unaccountably blurred images. There are two main causes of blurring. One is vibration as the shutter operates, possibly produced by the "slam" of the reflex mirror of an SLR camera as it moves quickly out of the light path, and the other is imperfect focusing. It can be surprisingly hard to focus even on the Moon – and if it is hard to get the Moon in focus, what chance do you have with Mars?

This is not the place to go into the solutions of these and other problems, but you must lick them before you can go on to photography at higher powers. This is achieved by means of *eyepiece projection* – increasing the effective focal length by using an eyepiece between the telescope and the camera to project a magnified image onto the film. By this means the *f*-ratio is increased considerably, maybe to several hundred, and exposure times become longer and longer. For the full Moon exposures may still be less than a second, but as you are magnifying the image more you will need a driven telescope. In extreme cases it must be driven at the lunar, rather than sidereal, rate.

The planets

Once you have mastered lunar photography at higher magnification you can proceed to the brighter planets. Your technique and ability will be tested to the limit to obtain anything like the amount of detail on, say,

Figure 6.7 *Saturn is only half the angular diameter of a crater such as Archimedes, with only a fraction of its brightness, and is correspondingly more difficult to photograph. A 250-mm (10-inch) reflector was used for this picture by Bob Garner.*

Jupiter that you can see visually. This is because the eye is able to register detail visible only in brief instants of good seeing. Film needs to be exposed for a large fraction of a second to record anything significantly above the chemical fog level of the film, so the fleeting detail that you can make out does not find its way onto the photographic image (see Figure 6.7).

If your results are poor to start with, you can console yourself with the thought that even a dark country site does not offer any advantage, as long as you are not badly affected by local industrial pollution. The limit is set by the seeing. The very best planetary photographs are taken from observatories where the seeing is considerably better than average – usually on mountaintops above an inversion layer (see Chapter 2). At low-level sites, the best seeing is often accompanied by mist. While visual observers can happily observe through all but the worst mist, it makes photography even more difficult than it already is.

Deep-sky objects

The exposure times needed for photographing deep-sky objects are much longer than for the Moon and planets. The telescope must not only be driven accurately, but also guided – that is, the drive rate must be monitored by observing a star and corrected as often as necessary to keep the star centred.

This allows you to use the same sort of exposure times as for piggyback or telephoto work, but with a larger image scale. Remember, however, that if your telescope is *f*/4 you will record sky background in just about the same exposure as you used with the *f*/4 telephoto lens, for a given film speed. The difference comes in the scale of the images. A planetary nebula such as the Ring Nebula (M57) in Lyra is tiny when photographed with a telephoto lens – it is barely distinguishable from a star. But a longer focal length magnifies it and shows it as a delicate red smoke ring (see Figure 6.8).

Figure 6.8 *Many planetary nebulae are bright enough to be observed from city skies. This photograph of the Ring Nebula (M57) was taken through a 250-mm (10-inch) reflector by Bob Garner from his home in West London, close by a brilliantly lit arterial road.*

Armed with a telescope on a driven equatorial mount and a means of guiding it, you can now attempt to photograph a wider range of objects than before. Again, establish your sky background limiting exposure and then find objects that are bright enough to register on the film before you reach that limit. It would help if deep-sky catalogues were designed more with the city observer in mind, but you will find that, as with visual observing, the term "bright" is misleading.

The magnitudes quoted in catalogues and guides are often photographic rather than visual magnitudes. In theory, they should be a better guide to your ability to photograph an object than they are to its visibility, but they are usually based on blue-sensitive emulsions and the results you get on panchromatic or colour film may be quite different.

Just as for visual work, do not stick to the Messier objects on the basis that they are the brightest in the sky. Many objects that Messier failed to include in his famous catalogue are well within the reach of your camera. Loose clusters in particular are worth attempting to photograph. And by using red-sensitive emulsions (covered in greater detail in the section "Black-and-white photography with filters" on page 117), you can easily record nebulae that are virtually impossible to see, even from dark sites and even without a filter.

Black-and-white or colour?

Speaking for myself, I like to use colour film. In fact, I hardly ever put a black-and-white film in the camera these days. For the city astronomer, however, there are advantages in using good old black-and-white. You can process it yourself (in fact you are almost obliged to, since most photo shops don't know what to do with it), and there is a wide range of special emulsions available. Another advantage peculiar to city astrophotography is that you can tolerate a higher sky-fog level. Whereas a colour photograph is spoiled for all practical purposes if the

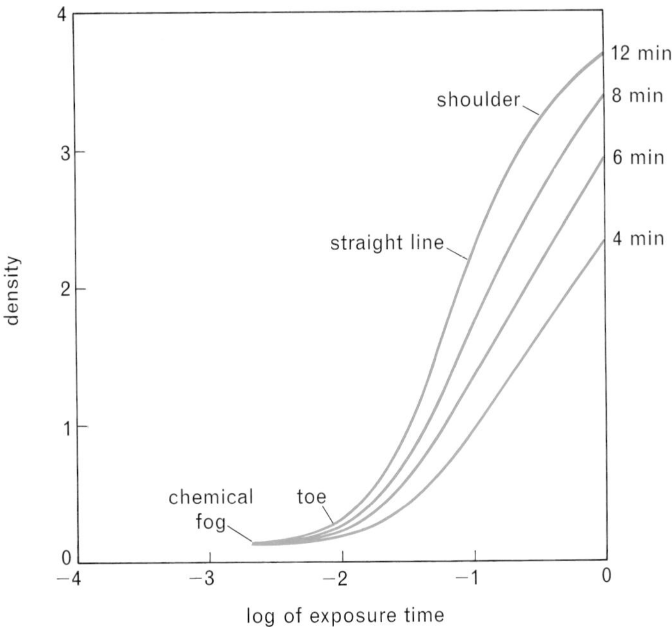

Figure 6.9 *A family of characteristic curves of Tech Pan emulsion, showing how the density of the processed film increases with exposure. The so-called chemical fog gives a background density before the film is exposed. The response to light is particularly poor at the low-exposure end, but past the toe of the curve, on the straight-line part, the film responds more quickly. Past the shoulder, the response approaches saturation level and further exposure will add little to the image. Increasing the developing time gives greater contrast (a steeper curve); for deep-sky work it is normal to give as full development as possible so as to bring out the full speed and contrast of the emulsion.*

background is yellow or green, a moderately dense fog level on a black-and-white negative is acceptable.

Astrophotographers actually set out to expose their film so as to get sky fog. This is because the contrast of the film emulsion is very low at low light levels. Just as when the contrast of a TV is turned down the image is hard to see, so at low contrast the faintest parts of a photographic image are barely discernible. The reason is obvious when you look at what is called the *characteristic curve* of the emulsion, which is simply a graph showing how the image density increases with the time of exposure (see Figure 6.9). It is easiest to consider a black-and-white film, but the same effect occurs with colour film.

You might think that, the more exposure you give, the greater the density of the image. But at low light levels this is not the case. The curve starts flat – that is, very dim objects have no effect on the emulsion at all. The density here is the *chemical fog* level, caused by the development process itself. The longer you develop even an unexposed film, the higher the chemical fog level. The curve then starts to lift, at what is called the "toe" of the curve. It then follows a "straight line" (which in practice is often slightly curved). The slope of the straight line is what is called the emulsion's *contrast* – the steeper the curve, the higher the contrast. With a high-contrast emulsion, a small increase in exposure produces a considerable increase in density.

We can now see why it is a good idea to have a certain amount of sky fog. If you time the exposure so that the sky fog just gets onto the straight-line part of the characteristic curve (that is, so that the negative looks quite dark even between the stars), the objects you are interested in – which are hopefully brighter than the sky fog – will be well onto the more contrasty part of the curve and will therefore be easier to see. It is established practice, for locations where even a long exposure does not produce an appreciable sky background, to *pre-flash* the film. This pre-exposes the emulsion with an even background, so that any additional exposure from astronomical objects is recorded straight away instead of having to climb past the toe first. In city skies this is done for you.

When you have used black-and-white film you can make a print which reduces the visual impact of this fog level somewhat, and you can choose contrasty paper on which to make the print. If this results in some parts of the image becoming washed out, you can apply a little darkroom trickery by waving a mask in front in the enlarger's light beam when making the print so as to give those parts less exposure. This is also possible when making colour prints, but the colour of the sky fog is unavoidable. You could filter it out during printing by turning it into a neutral grey, but the additional filter will also affect the colour of the objects you are photographing. So black-and-white film offers considerable advantages to the city astrophotographer.

Colour, on the other hand, can make all the difference to a photograph of a loose cluster by bringing out the star colours. And the red colour of nebulae makes even a dimly recorded image stand out, while in black and white the result would be of little interest.

Black-and-white photography with filters

Since sky glow usually has a noticeable colour, it is possible, to a certain extent, to filter it out when the photograph is being taken. There are indeed light-pollution reduction filters available, and I shall deal separately with their visual and photographic use. There is, however, a cheap way of filtering out light pollution: use a red filter. A Wratten 25 (often abbreviated as W25), 29, or 92 filter is often suggested for this purpose. The W25 is a standard photographic red filter, and has only a limited effect on sky glow. The W29 is a deep red, and as such transmits only a little visible light. The W92 is the deepest red of all, and gives a very dark view even in daylight. The deeper the red, the more light pollution the filter will cut out, but the longer the exposure that will be needed. The Wratten range of filters is made by Kodak and is available through good photographic dealers, though the particular one you want may have to be ordered. They are available quite cheaply as 75-mm (3-inch) gelatin squares, which must be handled with care. An even cheaper alternative is to use the plastic photographic filters from a "creative filter" range such as those produced by Cokin, though their numbers do not tally with those of the Wrattens and you will have to choose one by inspection. As a guide, the Cokin A003 is comparable to the Wratten 25. The material of which they are made is not optically worked to improve the flatness of the surfaces, and so they may distort the image. They are therefore probably adequate only for use with focal lengths shorter than 100 mm.

Glass versions are available from other manufacturers, who often use the same numbers as the Wratten filters. The Wratten 29 cuts out virtually all light to the blue side of 620 nm, but has a transmission of over 90 percent at the wavelength of the hydrogen-alpha line (see page 46). The Wratten 92 cuts out light to the blue side of 630 nm, but transmits rather less H-alpha. In addition, the US firm of Lumicon have their own range of glass H-alpha pass filters, for which they claim a 90 percent transmission at the wavelength of H-alpha, yet with a blue cutoff similar to that of a Wratten 92 filter.

The type of filter you buy may depend on the system you use for astrophotography. If you are using a telephoto lens, you can simply put the filter over the front of the lens. In SCTs there is a cell at the rear end which takes filters, but with other apparatus you may have to work out your own arrangement. If the filter is placed not over the front of the lens, but somewhere within the converging beam of light of the camera or telescope, the position of the focus will be affected. Be warned that it is extremely difficult to see light that has passed through a deep red filter.

A good way to use a gelatin filter is to cut a disk from the square so that it will fit inside a plain or UV filter that in turn fits into your lens. The material is too costly to be used at the aperture end of a telescope, so you will have to position the filter within the focusing mount. Gelatin is an animal product and is prone to attack by moisture. If it gets damp its performance will be degraded, and you will just have to buy another. One advantage of a gelatin filter over a glass one, however, is that the change in the position of the focus when it is used within the converging beam is negligible.

By day, a red filter such as a Wratten 25 increases the exposure by three stops, which is not extreme. Look through such a filter at a sodium streetlight, however, and you will find that it is greatly dimmed. This is an excellent means of reducing skyglow, though it is by no means perfect. First, your exposure times for star images will be increased considerably, because of the filtering action itself. Second, as a consequence, reciprocity failure (see below) could start to become bothersome. And third, the film itself may not be very sensitive to red light.

If the light from an object falling on the emulsion is so dim that long exposures are needed, the emulsion does not record the object very well. This is a result of what is called *reciprocity failure*. In everyday photography there is a reciprocity between the exposure time and the image brightness – that is, doubling one can be exactly compensated for by halving the other. However, at very low light levels this is no longer the case.

With the red filter in place during the day, your exposure meter will suggest that you increase the shutter speed from, say, 1/250 second to 1/30 second to compensate for the dimmer image. But when the light levels are already low, this no longer works. Increasing the exposure time from 2 to 16 minutes is not enough – you may have to go up to 30 minutes or more. With slow film in the first place, you could be facing very long exposures.

The colour sensitivity of film

The matter of the film's colour sensitivity is of great importance to astrophotographers, yet it is of scarcely any concern to everyday photographers. A basic emulsion is sensitive to blue light only. In old films and photographs you may notice that red flags appear black because the emulsions used had no red sensitivity. In both black-and-white and colour films, dyes are added to the emulsion to make it sensitive to other colours. These differ from film to film, so some films will perform much better than others when you use a red filter in front of them. Two in particular, Kodak Technical Pan (see Figure 6.10) and 2475 Recording Film, have very good red sensitivity. Tech Pan – formerly known as 2415, and still often referred to as such – is a particularly slow film, and 2475 is very fast indeed. The actual ISO speeds of these films depend very much on their development. Neither film is a stock item in ordinary photographic stores, but both should be obtainable from suppliers used by professional photographers. Kodak dealers can order them for you, but there is a minimum order quantity.

The crucial feature that makes these films so useful is that their red sensitivity extends to the H-alpha line. This spectral line is an absorption line in the Sun's spectrum, but it is a strong emission line in the spectrum of many gaseous nebulae. An object such as the North America Nebula, all but invisible to the naked eye except under the darkest conditions, shows up clearly on photographic film sensitive to H-alpha.

Figure 6.10 *Spectral sensitivity curves of Tech Pan black-and-white and Fujichrome 400 colour film compared. For the Fujichrome, separate curves are shown for each of the emulsion's colour-sensitive layers – blue, green, and red. Both films have a peak of sensitivity near 650 nm, coinciding with the strong 656.3 nm hydrogen-alpha line. The sensitivity curves of the Ektachrome family of films are very similar to those of the Fujichrome. Most black-and-white films are markedly less red-sensitive than Tech Pan, which gives them a better rendering of flesh tones for everyday photography.*

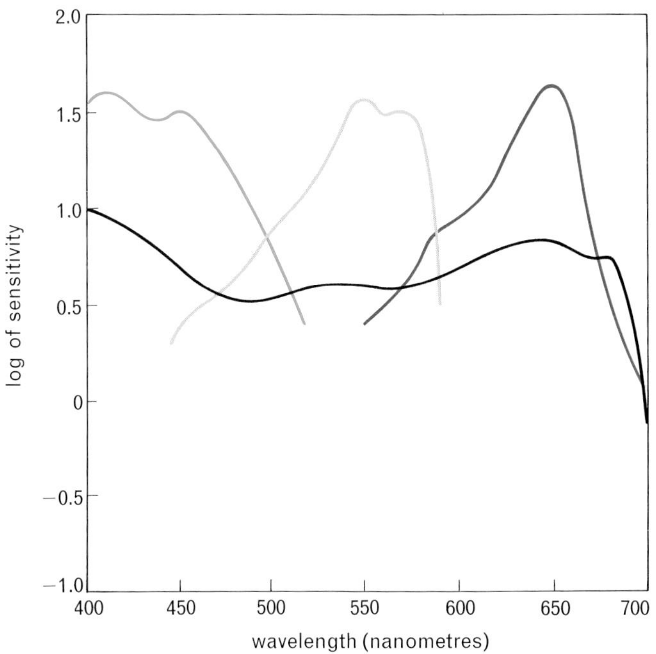

Its speed makes 2475 particularly useful when used in conjunction with a Wratten 29 filter. The drawback is that the graininess of the film is very noticeable, to the point of ugliness, which is inevitable for such a fast emulsion. Even from a badly light-polluted area you can easily photograph the Milky Way and numerous gaseous nebulae (that is, the so-called H II regions, which are predominantly glowing ionized hydrogen) with this film–filter combination. Its ability to provide dark skies is quite remarkable. Exposure times of 1 hour at *f*/2.8 are possible even from urban areas, though because of reciprocity failure the second 30 minutes will not add much to the results of the first 30 minutes. As well as the graininess of 2475 detracting from the pictorial appeal of your shots, in common with other fast films the image contrast can be rather low.

If you want to try a finer-grain film, there are more conventional emulsions which work fairly well, though they need considerably longer exposure times. Ilford HP5 Plus is one such, nominally rated at ISO 400 but usable at up to ISO 6400 with extended processing. Its Kodak equivalent, Tri-X, is not at all sensitive in the red. No conventional films have anything like the sensitivity to H-alpha of Technical Pan and 2475.

Hypersensitizing

The ultra-fine-grain Tech Pan is slow even by everyday standards, and used as it comes out of the box even its red sensitivity would not make it very useful to city astronomers. However, when hypersensitized – usually shortened to *hypered* – it performs extremely well because this process reduces the effects of reciprocity failure. In a test, David Cortner found that its speed for exposures of 1–3 minutes was similar to that of Kodak T-Max 3200. However, since Tech Pan has strong red sensitivity, it proved very much better when used with a red filter to cut down light pollution. T-Max 3200, by comparison, had almost zero sensitivity to H-alpha. Photographs on hypered Tech Pan by David Cortner and David Lehman are shown in Figures 6.11 to 6.13.

Hypering was originally achieved by slowly baking a film (with due precautions). It is now done by immersing the film in a gas which chemically reduces the emulsion – that is, it removes the water and oxygen. The classic way to achieve this is to bathe the film in hydrogen, but because of what is termed the "Hindenburg syndrome" this is now deemed too dangerous to use, even though the quantities of hydrogen required are quite small. Instead, forming gas is preferred, which is 8 percent hydrogen in 92 percent nitrogen. The gas is provided in cylinders, but it is not always easy to obtain a supply.

You can buy hypering kits, or you can buy your film ready-hypered from specialist suppliers. It has a limited shelf life in this form, so you will need to use it fairly quickly. If buying a hypered film does not usher in thirty successive nights of unbroken cloud, you can obtain some remarkably fine-grain and, with a red filter, light-pollution-free images.

This solution to the problem of light pollution provides another particularly suitable means of photographing H II regions. The combination also performs excellently on other objects. Even though stars and planetary nebulae do not emit strongly at the red end of the spectrum, the reduction in sky background means that even galaxies can be photographed. You will find that, although the filter cuts out only 10 percent of the light

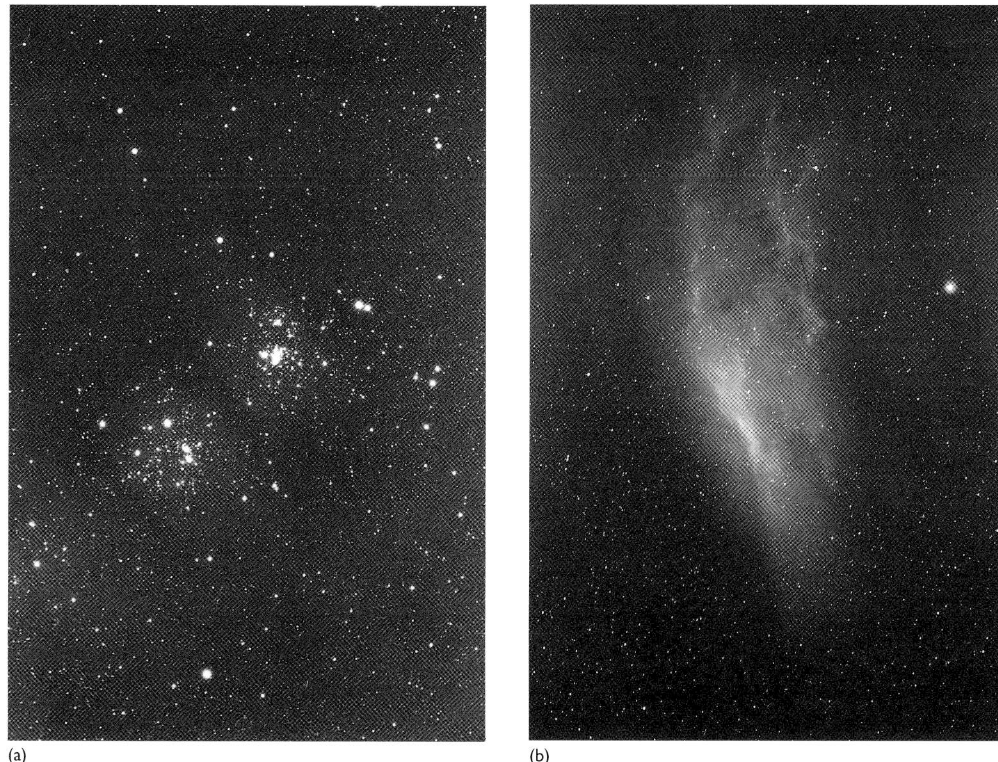
(a) (b)

Figure 6.11 *Two photographs on hypered Tech Pan film, taken by David Cortner from Johnson City, Tennessee. (a) The Double Cluster (NGC 869 and 884) in Perseus, taken using a 125-mm (5-inch) refractor at f/4.5 with no filter. One advantage of negative film is that corrections may be made during printing: the negative of this frame was so dense that nearly a minute was needed to expose the print, compared with a few seconds for a normally exposed subject. (b) The California Nebula (NGC 1499) in Perseus, taken with the same instrument but with a Lumicon H-alpha Pass Filter; exposure time 30 minutes. The negative of this picture was weakly exposed and required contrasty paper to bring out the detail.*

of H-alpha, your exposure times will have to be increased considerably compared with unfiltered exposures that show the nebula. The reason for this is that the filter cuts out most of the other nebular lines, which contribute around 65 percent of the brightness of a diffuse nebula, as well as much of the light pollution.

Using colour film

What about using the same technique with colour film? It has been done, but the problem is usually that the faster films in general have poor red sensitivity. This is because for everyday photography, red sensitivity tends to make the film very susceptible to the colour of lighting illuminating a scene. If the sunlight is slightly red, as in the late afternoon or in winter, it will have an undue effect on the red-sensitive layer of the film and will produce strongly reddened pictures. This would be excellent, however, for producing rich, red colours from H II regions.

The other problem is that using a red filter with colour film just turns everything red – stars, nebulae, the lot. Enterprising astrophotographers

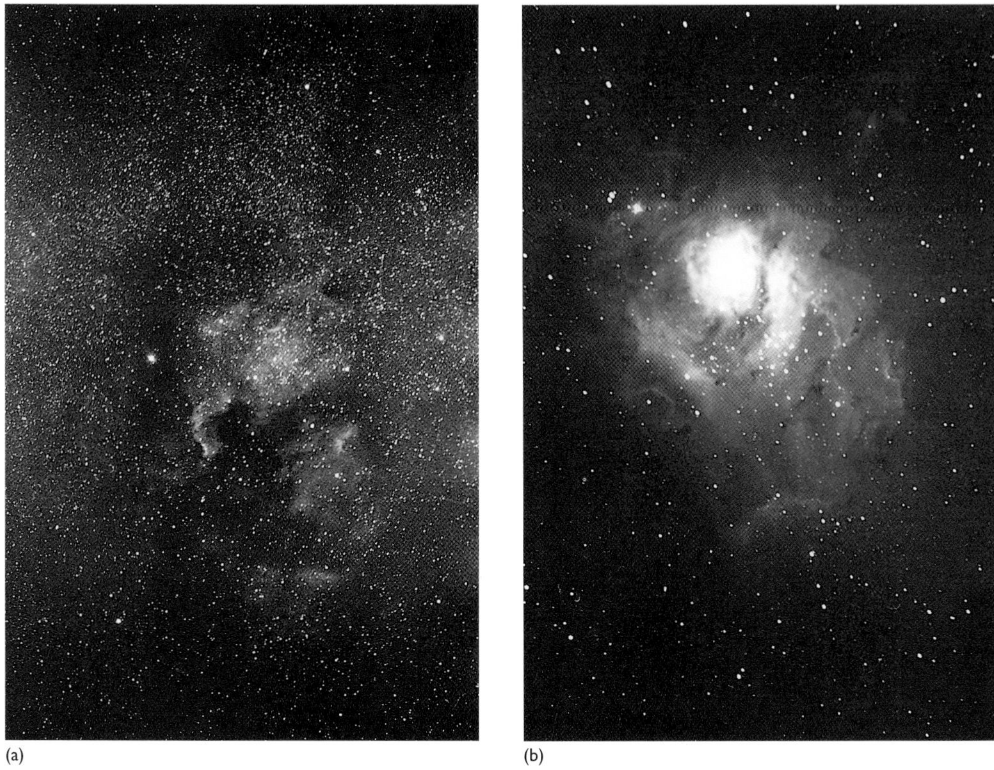

(a) (b)

Figure 6.12 *From his light-polluted backyard to the north of Fresno, California, David Lehman took these pictures of (a) the North America Nebula (NGC 7000) and (b) the Lagoon Nebula (M8) on hypered Tech Pan film using a 250-mm f/5 Newtonian reflector.*

have tried to overcome this by removing the red filter for part of the exposure so that the star images are burned in white. This requires a careful balancing act and a fair bit of trial and error to get the two exposure times exactly right. Mark J. Coco of Redondo Beach, California, used this method to take wide-field photographs published in *Sky & Telescope* (February 1986, page 215). He used a Cokin A003 filter with 3M Color Slide 1000 film. Although his skies were not as bad as typical city skies, his unfiltered 20-minute exposure of the Orion region at *f*/2.8 had a strong brown cast as a result of light pollution. He found that by using the filter over the lens for the first 10 minutes of the exposure, the sky background was darkened considerably while the nebulosity, such as Barnard's Loop and the Rosette Nebula, was virtually unaffected. Using a Cokin filter, which slips into a plastic mount fitted to the front of the camera lens, allowed him to remove the filter without disturbing the camera position – essential to avoid double images of the stars. From a city location the exposures would have to be considerably shorter, but the method bears promise and experimentation with different films and filters could yield very good results.

Some colour films are particularly poor at showing H II regions. The otherwise very useful Fujichrome 1600, for example, shows them as virtually white, if at all. The reason is that H II regions have other, weaker lines, in the blue and green regions, as well as the red line. A colour film also has three separate layers, photosensitive to red, green, and blue. The three visual emission lines of hydrogen gas record on their own respective film layers, with the result that each layer gets virtually the same amount of light. The inevitable result is a white nebula. The film's low contrast is a drawback here as well, making it rather unsuitable for photographing nebulae from the city. For stars, however, it is excellent as long as you can keep the exposure short enough to avoid excessive sky fog.

Colour films that are disliked by astrophotographers working from dark sites can come into their own in the city. I have found, for example, that Ektachrome P800/1600 exposed where there is light pollution has a tendency to give a blue background, which can give quite an attractive result. When I visited the well-known observer Robert McNaught in Australia, he expressed surprise that I was using the film, which he found gave unpleasantly harsh results. I, too, found that in the dark skies of Siding Spring it gave much less acceptable results than from the suburbs of London.

ScotchChrome 400 was found in a test published in *Astronomy* magazine (October 1992, pages 68–75) to have very good red sensitivity and a much better ability to reveal faint stars than any of the other films tested. It is rarely stocked by normal dealers, though, and you may need to find a specialist supplier. Outside the US you can also obtain ScotchChrome 800/3200P, which has the same red sensitivity but is designed for *push-processing*, in which the film is developed for longer to boost the contrast and effective speed. According to its manufacturer, 3M, this is a separate film, and not just an alternative name for ScotchChrome 400.

Manufacturers change the specifications of their films from time to time, and rarely with astrophotographers in mind. Sometimes their changes are fortuitously beneficial to astrophotographers, and sometimes the reverse. While published tests of films for astrophotography are usually carried out in good sky conditions, city astronomers have to devise their own tests, since the results may vary from area to area. Just because someone has pronounced a particular film good or bad, it does not mean that the same will be true for you. The nature of your local sky background could well affect how the film performs, possibly because of the effects of pre-flashing on the various film layers. The separate layers may have different reciprocity failure characteristics, leading to unexpected casts (colour biases). I once took a picture on Kodachrome of the Pyrenees from the Pic du Midi by dim moonlight. The picture came out deep red, presumably because the film's red layer had better long-exposure characteristics than the other two layers. Some films produce a lurid and unpleasant green sky background in the presence of light pollution.

Green or other coloured backgrounds can also be a characteristic of film that has not even been exposed to light. Look at the area around the sprocket holes or the rebates (the dark areas around each frame) to determine this. The background hue can also vary with the age of the film, and may even change after processing.

If you cannot escape the clutches of yellow light pollution on your colour slides, take a tip from Alan Heath of Nottingham. He has discovered that when making a contrasty black-and-white print from the colour slide, the yellow fog has little effect on the blue-sensitive paper emulsion. The worst effects of the light pollution therefore disappear,

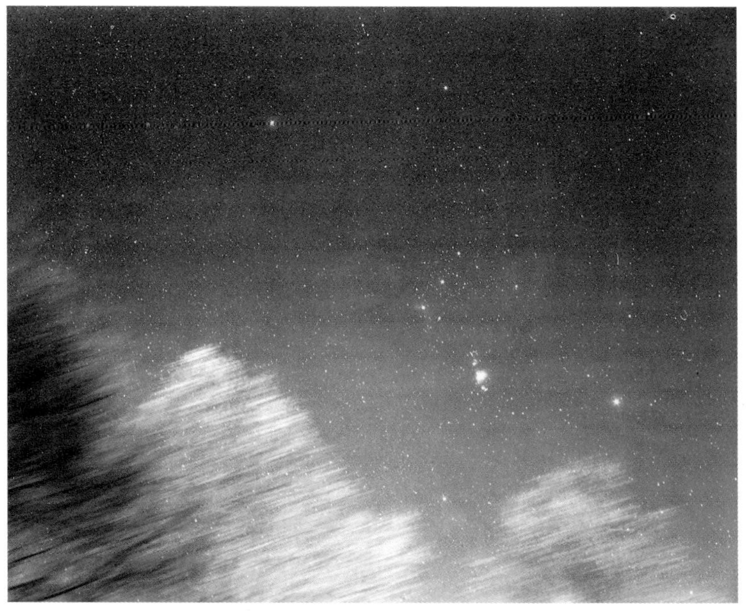

(a)

(b)

Figure 6.13 *A graphic illustration of the power of hypered Tech Pan film combined with an H-alpha filter. (a) The unfiltered view from David Cortner's backyard, in the direct line of sight of 42 sodium and mercury lamps, some of which are visible at the bottom of the frame. Exposure time 5 minutes. (b) The same view, but exposed for 30 minutes with an H-alpha filter. The Rosette Nebula at the left of the frame and the long Barnard's Loop, curving its way around Orion, are clearly brought out. Just visible below Orion's belt is the nebulosity (IC 434) containing the Horsehead Nebula.*

leaving just the star images. The technique is equivalent to using a fast, blue-sensitive black-and-white film, or a panchromatic film with a blue filter. Its drawback is that the interesting H-alpha sources will disappear along with the light pollution.

Processing

Most people probably send their colour film to a lab for processing, either directly or through a photo shop. If you are using print film, there is a further source of background casts – the print machine. Unlike slide film, print films are not manufactured with a fixed colour balance. Whereas a slide film that gave, say, predominantly blue pictures would be very unpopular, when prints are being made it is a simple matter to alter the colour balance by using filters so as to allow not only for differences in film manufacture, even from batch to batch, but also for differences in lighting conditions. This is why the same print film will take pictures in all types of lighting conditions, whereas you need one type of slide film for taking pictures in daylight, and another type for interiors lit by tungsten lighting.

When making a colour print, the use of filters for colour balancing has to be carefully controlled. This makes colour printing by hand a very time-consuming affair, but the machines used for processing rolls of colour prints have automated colour analysers. These look at the overall colour balance of the film (after subtracting the orange mask which is present in all print films), then assume that there are the same proportions of all colours. For everyday purposes they can be quite accurate, though they tend to give washed-out results on pictures which consist of a predominant single colour, notably sunsets.

One often reads in astronomy magazines of people's success in obtaining prints from 1-hour photo shops, but many others have had bad experiences. Not only will the processing machine wash out the colour, it will try to force the background to an overall grey. If the background was really sodium yellow, this could have strange effects on the colours of the stars and nebulae. Furthermore, the background usually comes out a much lighter shade than one would like. Everyday pictures are rarely predominantly black, and print machines are not designed to cope with pictures of the night sky. Very often the machine will refuse to print the picture at all, which is particularly galling if you get charged by the number of shots on the film, rather than by the number of prints you receive. If it does print them, you may get them back with a sticker telling you there is something wrong with your camera.

Leaving a note with the film to make prints regardless of the content may do the trick if the technician is alert, but if the film is sent away for processing the note will probably be overlooked. Your best bet is to go somewhere where films are processed on the premises, and to speak to whoever will be supervising the machine when your prints go through. They may well be able to override the automatic mechanism, though do not expect the same quality as from hand prints.

The other problem with lab processing, less commonly noticed with photographs of city skies, is that it can be difficult for the lab or the machine to tell where the individual frames on the film begin and end. You could well get frames cut right through the middle. This applies to slide films as well as to prints. One solution is to make sure that you have a daytime or at least a well-exposed shot at either end of the film, so that whichever end the machine or operator works from there is something to show where the frames are. The frames are generally at a standard interval, which ties in with the frame numbering which is put on the film by the manufacturer (though the numbers may not match your own frame numbers). If you are using slide film, you could ask for it to be returned unmounted so that you can do your own mounting afterwards.

Image-processing software packages designed for the computer manipulation of CCD images are now widely available. They enable remarkable amounts of information to be extracted from raw images (see Figure 6.20), and many of the benefits of using CCDs derive from what it is possible to do with the software. Although the results of image processing can seem magical, it is simply a matter of taking information that is already there and making it more visible. For example, it may be that certain details in the image are hidden because the brightness range that the monitor can display is limited. You could reveal such details by altering the brightness and contrast of the display, but then other details are will become too bright or too faint to be seen. A process called logarithmic rescaling alters the brightness values of the image, enabling a wider range of detail to be displayed. Another trick, borrowed from conventional photography, is unsharp masking: this increases the contrast of small-scale detail while suppressing brightness differences on a larger scale.

Photographers have not been slow to appreciate the benefits of applying digital processing to conventional photographic images. First, however, the images must be scanned to convert them to digital form. Hand-held scanners are available quite cheaply for scanning prints at comparatively low resolution, but if your originals are in the form of transparencies you must either make prints and scan those, or have the transparencies scanned by a bureau. The Kodak Photo-CD system will scan slides or negatives at very high resolution onto a compact disc.

Another use for computer processing is in combining separate images. This enables you, for example, to take three black-and-white images of an object, one each through red, green, and blue colour filters – either using a CCD, or, for optimum quality, on hypered Tech Pan film – then combine them to make a colour image. On a computer the task of keeping the images in registration is simple, but the file sizes for good quality images from scanned photographs can be enormous, and an advanced machine with good graphics handling is necessary. Finally, the image must be converted to a hard copy, such as a slide.

Countering light pollution with deep-sky filters

It is not often that a miraculous cure is invented, but the filters now available for deep-sky work could come into that category. They do indeed work wonders – but only in the right circumstances and on the right objects. Regrettably, they are not the complete answer to the light-polluted amateur's prayer. You could spend a considerable sum of money and still find that you cannot see or photograph what you were hoping to.

The filters are designed to screw into the barrel of standard sizes of eyepiece, onto telephoto lenses, or into a cell specially provided on SCTs.

Their aim is to cut out radiation from those parts of the spectrum where the streetlights produce their light. The simplest example, and these days the least useful, is the *didymium filter*. This is a glass filter which absorbs light at exactly the same wavelength as that emitted by a low-pressure sodium streetlight. Its effect is truly remarkable. When viewed with white light the filter looks fairly clear – it has a noticeable grey-blue colour, but if you had sunglasses made of didymium glass you would say they were tinted rather than dark. Look at a sodium streetlight, however, and its light all but disappears. It has a high absorption at the main wavelength emitted by a sodium streetlight, but transmits virtually every other wavelength.

In a perfect world, all outdoor lights would be low-pressure sodium and all astronomers would simply equip themselves with didymium filters at the focus of their telescopes. This is the position taken by professional astronomers in the US, who hope to persuade the authorities in cities near observatories to use only low-pressure sodium streetlighting.

My own experience with didymium filters is not very encouraging. I obtained one in the early 1980s, when the majority of streetlights in my area were still low-pressure sodium. While it performed impressively on individual streetlights, I found that even on a bright object such as the Orion Nebula the view was not improved by using one: although the sky background was darkened, so was the nebula. Using my camera's exposure meter, I found that the filter required an increase in exposure of one stop – that is, it had the effect of halving the brightness and turning my 220-mm (8½-inch) telescope into a 150-mm (6-inch) one. Now, a view of the Orion Nebula through a 150-mm telescope under good conditions is impressive, but I was less than impressed with my view. I can only conclude that, even then, there were a significant number of non-sodium lights around, the effects of which were not removed by the didymium filter. Today, with a higher proportion of high-pressure sodium lighting, the filter is of little use. Photographically, I found that pictures taken through the filter were no better than those without it. But under the right circumstances such filters could be very useful – if, for example, the only nearby illumination were low-pressure sodium.

Interference filters

The didymium filter works in the same way as an ordinary red or yellow filter – that is, by absorbing a certain part of the spectrum. But the filters which have made the most difference to observers work in a different way: they use the interference between different wavelengths of light to cancel out specific wavelengths. The precise mechanism is given in detail in physics textbooks, but basically it is very similar to the way oil on a wet road reflects certain colours, creating the familiar rainbow appearance. These filters, sometimes called *multi-layer filters*, have layers coated as thin as the wavelength of light, and can be made to transmit chosen wavelengths.

Interference filters are immediately obvious from their appearance. Whereas filters that absorb light look the same colour no matter which way you look at them, interference filters are a different colour depending on whether you look through them or at light reflected from them. The colours that are not transmitted are reflected away. So a filter that transmits blue light, for example, reflects the remainder, which is yellow.

Furthermore, the exact wavelengths transmitted or reflected depend on the angle at which they are viewed. They are designed to be used with the light passing through at right angles to the surface, and thus produce the best results when the object being viewed is exactly on-axis, at the centre of the field of view.

If your field of view is narrow, as at high power, this requirement is usually met. But with wider fields of view more light arrives at an angle. This results in stars near the edge of the field of view appearing a different colour from those near the centre, which can be disconcerting. When used with wide-field designs of eyepiece, such as Naglers, multi-layer filters will give their best results only at the centre of the field of view.

Deep-sky filters, which are also generally known as *light-pollution reduction* (LPR) filters, fall into two broad categories. Some cut out wide bands of the spectrum in which the main light-pollution lines are to be found, and therefore transmit quite a large amount of light from objects with continuous spectra, such as stars and galaxies. Others, which are more expensive, cut out the whole spectrum except for a band which is chosen to include a particular set of emission lines from nebulae. These are called *nebular filters* or *ultra-high-contrast filters*. The abbreviation UHC, often used collectively for these filters, is a tradename of Lumicon, one of the leading suppliers. Within this group there are narrowband filters which isolate specific lines.

Broadband filters

The lowest-priced multi-layer filters block most green and yellow parts of the spectrum. The Lumicon Deep-Sky Filter™ is one of these, and the granting of trademark status to it has required other manufacturers to choose other names for their own. Examples are the Orion Telescope Center's SkyGlow™ Broadband Filters and the Meade Series 4000 Broadband Nebular Filters.

There is one very unpleasant fact of life where filters are concerned. We have evolved on a planet orbiting a completely average star, and our eyes are naturally most sensitive to the peak of its radiation, which is yellow light. This is why the artificial light we tend to prefer, for streetlighting and other purposes, is predominantly yellow. Since other galaxies and clusters consist, on the whole, of other average stars, their light is roughly the same colour as many of our streetlights. So by cutting out the light from a streetlight, particularly a high-pressure sodium one, we are also cutting out a major part of the light of a distant star cluster or galaxy.

This means that no light-pollution filter can be totally satisfactory for viewing the majority of deep-sky objects, which are either galaxies or clusters. If you are fortunate enough to live where most of the streetlights are blocked out by simple LPR filters, the contrast of deep-sky objects against the background can be improved. The effect will be to reduce the aperture of your telescope, so you should be prepared to take every possible step to increase contrast rather than relying on the filter to wipe out light pollution.

Nick Hewitt, Director of the BAA's Deep Sky Section, uses a Lumicon Deep-Sky Filter from his home in Northampton, England. This town has spread enormously since the early 1970s, and housing estates now sprawl over what not long ago was pleasant countryside little changed since the eighteenth century. Nick says that while the filter does not actually enhance a galaxy, the increased contrast helps you to see it against the background. While it helps you to locate, say, the Whirlpool Galaxy (M51), it does not allow you to pick out its spiral arms.

On the other hand, Howard Brown-Greaves, who lives in Fulham in London, just 6 km (4 miles) from the bright lights of the West End, finds that his Lumicon Deep-Sky Filter is of no use in locating objects using a 220-mm (8½-inch) *f*/6 reflector in his very badly light-polluted skies. I have tried a Kenko filter of this type and found that I could see objects just as well without it, with the possible exception of the galaxy NGC 205 (near M31) when it was virtually overhead, through my 220-mm *f*/8 reflector. Even on NGC 205 the effect was marginal. As with the didymium filter, the effect on an individual streetlight was marked, but on the sky it was insignificant. David Cortner agrees that LPR filters "hardly help at all with these objects."

So is it worth buying a general broadband light-pollution filter? In suburban areas in the UK there is a large proportion of low-pressure sodium streetlighting, giving the sky a noticeable yellow glow which will be significantly removed by the Deep-Sky Filter. Near major city centres, however, there is a higher proportion of high-pressure sodium and other types of lighting, such as fluorescent and tungsten. Lumicon advertisements claim that their Deep-Sky Filter "improves your view of *all* deep-sky objects including galaxies and comets. It removes natural and man-made light pollution and yields 3–6× contrast gain for nebulae and 1–3× contrast gain for clusters and galaxies." Could the range in performance relate to the type of light pollution present? I put this question to the proprietor of Lumicon, Jack Marling, who responded that:

> the various LPR filters help, no matter what streetlights are in use .... For the Deep-Sky Filter, the 1–3× refers to no improvement (1×) to threefold contrast gain, primarily for photography of nebulae. For galaxies there is a 1–1½× contrast gain. The Deep-Sky Filter will help visibility of face-on spiral galaxies like M101, M33, etc.

There is no guarantee that a broadband filter will be of any use for visual observation of galaxies and clusters: 1× improvement. The inevitable reduction in brightness which it gives could actually make the object you are searching for more difficult to see. Nevertheless, the urban astronomer needs every bit of help, and the light reduction is not too significant if you use a fairly large telescope in the first place. This is where the large Dobsonians come into their own.

Nebular filters

Undoubtedly the most useful multi-layer filters are those which cut out virtually all light except for a relatively small band that includes a few emission lines. Examples include the Lumicon UHC Filter, the Orion Telescope Center SkyGlow UltraBlock Filter, the Daystar 300 Filter, the Meade Series 4000 Narrowband Filters, and the Orion Optics (UK) EHC Filter. These make use of the fact that most gaseous nebulae (with the exception of reflection nebulae, which appear blue on photographs) emit bright-line rather than continuous spectra. By viewing them in the light of these lines only, you can in theory observe the light from the nebula, and cut out the rest (see Figure 6.14). This removes almost completely the light from low-pressure sodium and mercury, and a large part of tungsten and fluorescent. High-pressure sodium has a line very close to the important O III lines (produced by doubly ionized oxygen), so it is not reduced as much as one would like.

A typical nebular filter has a bandpass (width of transmission band) of some 25 nm, which includes the green lines of H-beta and O III. This allows you to view most H II regions and planetary nebulae. Other filters are available that transmit an even narrower band, so that only the light of either H-beta or O III gets through.

A filter such as the Lumicon UHC is ideally suited to observing emission nebulae in whose spectra are lines of hydrogen, oxygen, or nitrogen. This includes H II regions such as the Lagoon Nebula (M8) and the Omega or Swan Nebula (M17), both in Sagittarius. It is also good on most planetary nebulae, such as the Owl Nebula (M97) in Ursa Major, and on supernova remnants such as the Crab Nebula (M1) in Taurus, which can be surprisingly hard to see from the city on account of its fairly low surface brightness. These filters are not likely to be of any use at all when viewing star clusters, galaxies, or the blue reflection nebulae.

The H-beta filter allows through only a 9-nm band centred on the green H-beta line at 485.6 nm. Oddly enough, it is best for very faint objects that appear reddish-pink in photographs, such as the elusive nebula IC 434 surrounding the Horsehead in Orion, or the California Nebula (IC 1499) in Perseus. Their pink colour is a combination of H-alpha, which registers on the red emulsion of a colour film, the weaker H-beta, which registers on the green, and H-gamma, on the blue. The eye is most sensitive to the green H-beta line, and scarcely at all to the red. This filter works best on objects that give out only hydrogen emission, and show virtually no other lines, and is therefore most appropriate for nebulae which are glowing under the influence of a star some distance away, so that only hydrogen, which is readily excited, is glowing. It is not particularly effective on most planetary nebulae, where the star that is causing the nebula to glow is closer and therefore able to excite oxygen and nitrogen too. Some planetaries, however, have a low-temperature central star which is not able to generate the O III lines. An example is NGC 40 in Cepheus, a magnitude 10.5 object about half the size of M76. These filters, then, have a very limited use and are probably most suitable for the dedicated deep-sky observer. It is also quite likely that from badly light-polluted skies even an H-beta filter will not show the faint nebulae. Lumicon claim that you will be able to see the Horsehead Nebula as long as the sky is dark enough for you to easily see the (northern hemisphere) winter Milky Way.

For viewing most planetary nebulae an O III filter is more appropriate, as their output of this green line is stronger than their hydrogen lines. Nick Hewitt finds that the notoriously large but low-surface-brightness Helix Nebula (NGC 7293) in Aquarius is "pretty easy with an O III filter" in skies in which it is otherwise invisible. He can also see the Veil Nebula (NGC 6992) in Cygnus by using a Lumicon UHC filter in his 200-mm (8-inch) Celestron SCT. For years this object was regarded as one of the most challenging for the deep-sky observer, even from dark sites, but it is

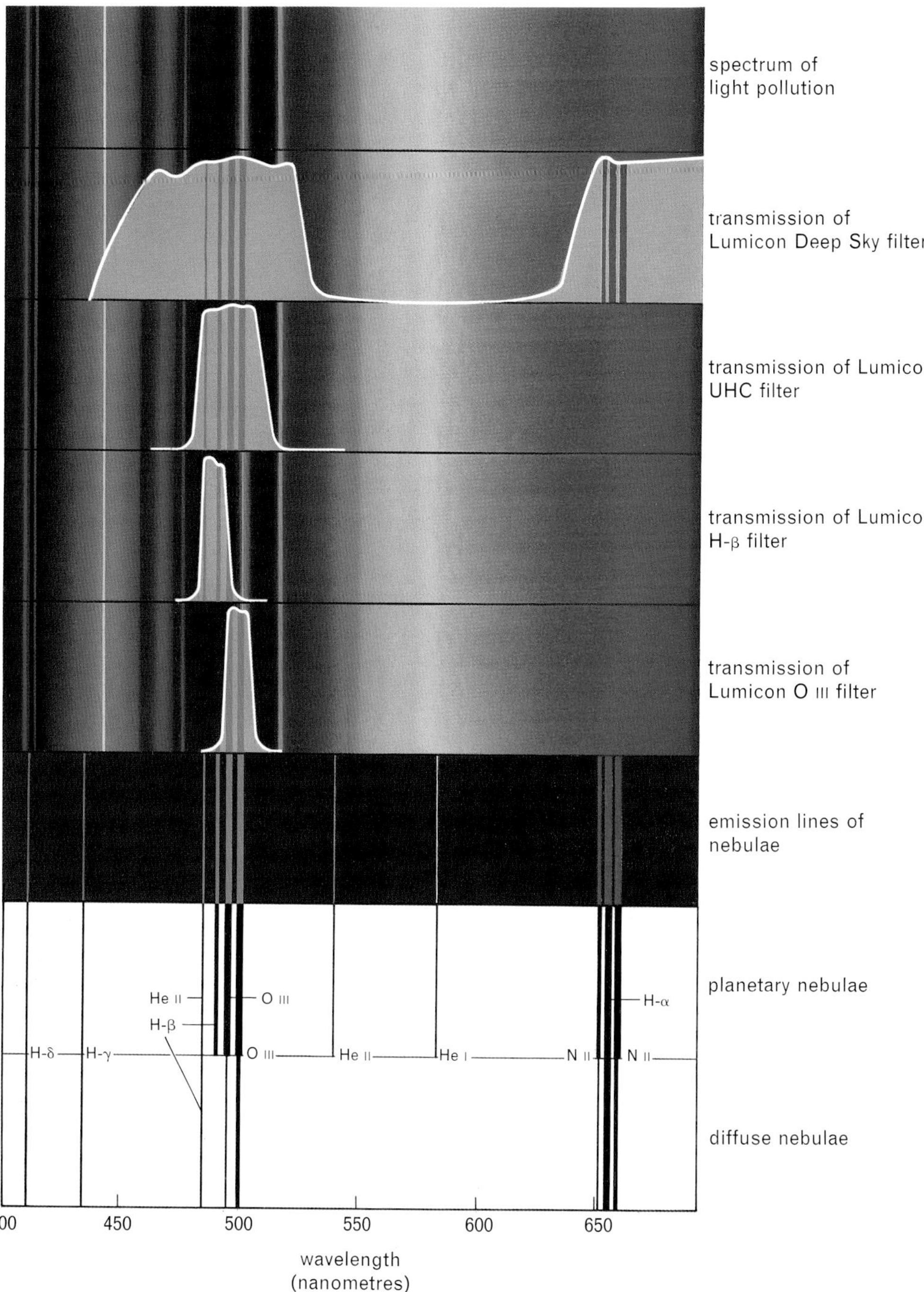

Figure 6.14 *Transmission curves of Lumicon filters, compared with a typical spectrum of light pollution (mostly high-pressure sodium, with some mercury) and important nebular lines. In each case the transmission of the filter is shown as a window which admits particular lines from nebulae while excluding most light pollution. The light-pollution line at around 500 nm, caused by high-pressure sodium, is very close to the important O III lines in the spectra of planetary nebulae.*

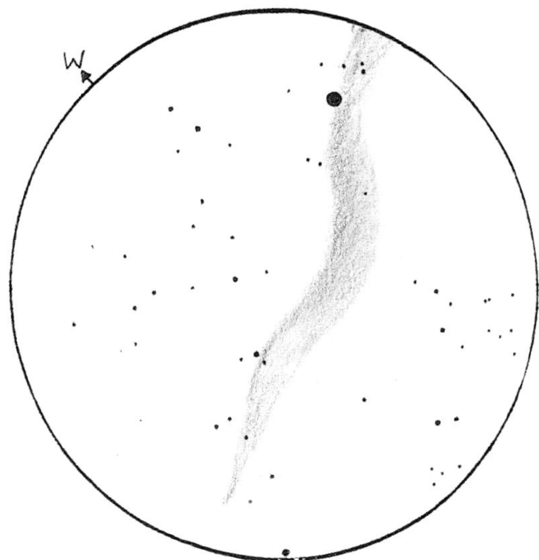

Figure 6.15 *Interference filters can reveal certain objects to visual observers. Here, Darren Bushnall of Hartlepool, England has drawn his view of part of the Veil nebula as seen through an O III filter used with a 220-mm (8½-inch) reflector.*

now visible with little difficulty from all but the worst locations, despite the great increase in streetlighting, thanks to ultra-high-contrast filters (see Figure 6.15).

Owen Brazell of the Webb Society and the BAA's Deep Sky Section has considerable experience of using filters from a variety of sites in Britain and North America. He feels that all but the hard core of deep-sky observers will find the nebular filters such as the UHC more suitable than narrowband filters. "Because the filter passband is relatively wide, the attenuation of starlight is not too great and the views of nebulosity associated with stars (such as M8 and M46) are very pleasing," he reports. They will help cut down the sky background even from heavily light-polluted neighbourhoods and allow you to see objects that would otherwise have been at the threshold of vision. He warns, however, that the narrowband filters have very restricted uses:

> The view of M42, for instance, through an O III filter is, quite frankly, disappointing. But if your interest is in hunting out faint and obscure planetary nebulae then this filter is a must. These objects will simply be invisible without it. The detail which can be brought out in nebulosity with this filter is incredible.

Star images are dimmed by about five magnitudes, so some of the aesthetic appeal of a nebula in a good starfield is lost. The O III filter works on most planetaries, but not on the Crab Nebula, whose light is produced by a different mechanism – synchrotron radiation.

Owen also reports that you can make interesting comparisons of the views of bright objects such as the Orion Nebula through O III and H-beta filters. The O III filter shows detail only in the central regions of the nebula, where the ionization is strongest, while the H-beta will reveal detail in the outer parts that cannot be discerned by any other method.

David Cortner, observing from Tennessee, says that on bright nebulae:

> a good LPR filter (like Lumicon's UHC or Orion's UltraBlock) can help salvage most of what streetlights take away. M16, M17, M8, and M20 are all helped immeasurably by the UltraBlock filter; M42 doesn't need the filter's help; and fainter nebulae walk a fine line of invisibility. The filter increases contrast but at some cost in brightness – if the latter effect takes faint nebulae below the limit, the increase in contrast is worthless. Faint nebulae in Auriga and Cassiopeia fall into this category. Lesser nebulae in Orion (east of Zeta, north of M42) are sometimes helped, sometimes hindered by the filter, depending on sky transparency and brightness.

As for planetary nebulae, David's comment is that:

> What filters can do for bright nebulae, they can do for planetary nebulae in spades. Besides, some planetaries have such extraordinary surface brightness that they shine through the glowing sky without ill effect. These also bear up well under high magnification (which, in turn, darkens the sky "foreground"). The Saturn Nebula, M57, M27, the Eskimo, all shine nicely for me. The Helix is visible on the very best nights from here, despite its low elevation and its position over the worst of my city's glow. It is huge, of course, and to see it at all I need the LPR and the 16-inch [400-mm] Dob working with a wide-angle eyepiece. The Veil can be beautiful when near the zenith and with the LPR filter. It's easy to see in the Dob with the filter and often shows good filamentary structure – both arcs are usually easy, and on the best nights I can even see the brightest knots in the middle. Only the brightest portions of the two bright arcs appear without the filter, and then without any detail. In the [125-mm/5-inch] refractor, the Veil is utterly invisible without the filter from here, and only dimly so with it. But as is so often the case, the very wide view in the refractor and the diamond-hard stars add a "context" to the Veil which makes up for much of the brightness lost in the smaller glass.

Nebular filters are still worth taking with you on trips to dark sites. From the caldera on Tenerife, at an altitude of around 2000 metres (7000 feet), I had a good view of the Veil through a 220-mm (8½-inch) reflector equipped with a Lumicon UHC filter. Without the filter, the Veil was considerably less easy to see. As mentioned in Chapter 7, the skies of Tenerife, while excellent by city standards, do suffer from the lights of coastal resorts. In addition there is natural skyglow, which means that the sky between the stars is never completely black.

For comet observers, Lumicon offer their Swan Band Filter. It transmits a 25-nm band which includes the green lines due to carbon at 514 nm emitted by some comets, notably those which have a high gas content rather than dust. This is fine for visual use on some comets, but it does not work for them all, and nor is it suitable for photographing comets.

It is also possible to obtain multi-layer filters which cut out just the yellow light from a low-pressure sodium lamp, in the same way as a didymium filter but more precisely. These are similarly useful where your main source of light pollution is low-pressure sodium, but they cannot combat high-pressure sodium light. In fact, the spectrum of high-pressure sodium dips at the very wavelength at which a low-pressure sodium lamp emits most of its light. The Natrium Filter from Orion Optics in the UK is of this type.

Photography with deep-sky filters

You can obtain deep-sky filters in sizes large enough to screw onto your telephoto camera lens. You can then take photographs of the larger nebulae, such as the North America and California nebulae, even from badly light-polluted areas. This is possible because the transmission of these filters increases sharply at the red end of the spectrum, like H-alpha filters. Nick Hewitt has used a Lumicon Deep-Sky Filter for this purpose and is pleased with the results (see Figure 6.16). "It is very good value," he says. "However, on colour film you get a brown background and some funny star colours." One of the objects he has photographed with this filter is the elusive supernova remnant IC 443 in Gemini, described in *NGC 2000.0* as "faint, narrow, curved."

David Lehman of Fresno, California, has a limiting magnitude of 4.5 to 5.0 from his backyard. He uses Technical Pan film and a deep-sky filter to take pictures through his 250-mm (10-inch) *f*/5 reflector. Among the subjects he has successfully photographed are the Cocoon Nebula (IC 5146) and the Veil Nebula, both in Cygnus, and the galaxy NGC 7331 in Pegasus. On this photograph he could even see the nearby string of 15th-magnitude galaxies known as Stephan's Quintet.

In general, the narrowband filters are not recommended for photography, since the response of film is somewhat different from that of the eye. Most films show a significant dip of sensitivity in green light, which is what the filters are designed to transmit. Exposure times will inevitably be quite long, particularly if your system is rather slow. Lumicon's Minus Violet filters are intended for use with colour film. They are not designed to reduce light pollution, but to cut out blue light, which some camera lenses do not bring to the same focus as the other colours.

The CCD revolution

If narrowband filters have worked a miracle, then the advent of CCDs – *charge-coupled devices* – has worked an even greater one. It is no exaggeration to say that these tiny microchips have reclaimed observational astronomy for the city amateur. Objects which became invisible in city skies many years ago can now be observed again as a matter of routine. There is hardly a class of object which cannot be imaged better with a CCD.

As always, there is a downside, and here it is in the cost and level of technical sophistication. But astronomy can be a technical subject, and much the same arguments can be levelled at the use of photography. One diehard visual observer has commented that using CCDs to find objects is "cheating" – to which the CCD convert justifiably replied, "So is using a telescope!"

A CCD is essentially an image-recording device which produces an electrical output. There is nothing unusual about the ones amateur astronomers use. You will find a CCD behind the lens of virtually every camcorder. The simplest way to obtain an electronic image through the

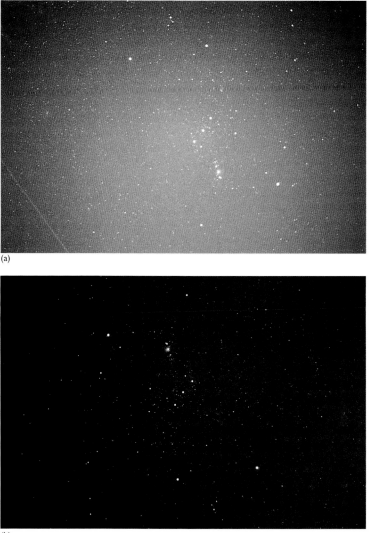

(a)

(b)

Figure 6.16 *Orion, photographed by Nick Hewitt of Northampton on Agfachrome 1000 film. (a) A 1-minute exposure using a 50-mm lens at f/1.8, and (b) a 2-minute exposure through a Lumicon Deep Sky Filter. In (b) the colour of the Orion Nebula is reproduced more accurately, and visible on the original transparency is Barnard's Loop, a large circular structure to the east of Orion, along with the nebulae IC 434 and NGC 2024.*

telescope is to point a modern camcorder into the eyepiece. A standard 5-lux camcorder will show detail on Jupiter, for example, which it is quite hard to capture on film using the same telescope. The Moon can be almost as stunning viewed in this way on the TV screen as it is to the eye – though it has to be said that the electronic scene always loses out in sheer detail and clarity to a directly viewed image. When it comes to stars, however, only the brightest can be picked up on a camcorder in this way.

The domestic camcorder is designed to operate in conjunction with a TV set, so each image is recorded at the normal TV image rate of

1/30 second (America and Japan) or 1/25 second (everywhere else). Just as with photography, increasing the exposure time for each frame records fainter images. The CCD systems used for astronomy do just that.

While camcorders are intended to produce their output for recording on videotape, this is unsuitable for astronomical images which may be produced only every few seconds or even minutes at the end of an exposure. Astronomers need to have access to their images in a permanent form, so although it is perfectly feasible to send the image from a CCD to a video monitor, it is necessary to record the image in digital, rather than analogue form. Consequently, most CCDs are designed to be linked to a computer.

At the heart of the standard CCD set-up is the image-recording chip itself, which is often only a few millimetres across (see Figure 6.17). This is mounted at the focus of your telescope, in place of the eyepiece or a conventional camera. Although professional astronomers use CCDs which have been developed for astronomy, those for amateur use are off-the-shelf chips intended originally for medical imaging or broadcast television. The chip used in the popular ST-4, Lynxx, and Electrim CCD cameras, for example, consists of an array of individual picture imaging elements, called *pixels*, in a 192×165 rectangle. This is a considerably coarser array than is used for chips in domestic camcorders. Other systems use larger chips, such as the ST-6 with its 375×242 array, and the Starlight Xpress, with 512×256.

To improve the chip's performance, it is normally cooled to about 50°C (90°F) below the ambient temperature. This can be achieved by Peltier cooling, which is an electronic method. When an electric current is applied to a Peltier cooling device, one end gets cold and the heat extracted appears at the other end. Chips cooled by this method may therefore be fitted with fins to dissipate the waste heat. Some systems use water cooling to prevent the heat from being released into the optical path, and to reduce the chip temperature still further.

A CCD camera also includes a control unit which links the system to the user's computer. There are variants on the method used to operate the chip, each with its own adherents. I shall not go into the various systems available, since new devices come onto the market regularly. The cheapest units cost only about as much as a basic SLR camera, while the more advanced ones can cost as much as a computer-controlled telescope. As well as the CCD system, you need the computer itself, which may well cost about the same again. You will have to operate your computer near the telescope, which may be undesirable if you have no observatory. Even if you use a battery-operated computer so as to avoid high voltages, kept permanently set up on a trolley which you can wheel out when needed, you will still need to protect it from outdoor conditions. Unless you are lucky enough to live in a year-round mild climate, these conditions can range from dew heavy enough to soak everything in sight to bitter frost. Neither are compatible with computers, so if you cannot run the computer equipment from a shed very close to the telescope, you will need a portable enclosure that will provide the right environment.

The additional, hidden cost of using a CCD is the time it takes to learn how to operate one successfully. If you are not used to computer technology,

(a) (b)

Figure 6.17 *(a) The head of a Starlight Xpress CCD unit. The chip itself, which records the image, is the small grey rectangle at the centre and measures 6.4×4.5 mm. The housing includes an electronic cooling device to reduce the "dark current" that occurs even in the absence of light. The fins at the back of the unit dissipate waste heat from the cooler. The head weighs 700 g (25 oz). (b) Four CCD chips compared. Left to right: the Texas Instruments TC211, as used in the popular ST-4 unit, with frame just 2.64 mm square; the Sony ICX027, used in the Starlight Xpress; the Philips NXA1011, a 600×284 pixel video CCD with high sensitivity, often used for astronomical work; and the CCD002, made by EEV for professional astronomers, with 576×384 pixels.*

there is also the matter of getting the hang of such intricacies as operating systems and disk management. Learning to run both a computer and a CCD together is a daunting task.

With all these drawbacks, the advantages of CCDs have to be pretty convincing to make it all worthwhile. Fortunately, they are. There are several features of CCD imaging that makes it superior, in its own way, to photography, and especially so from the city. First, while a photographic emulsion will continue to store light for some time after it is first exposed, as mentioned earlier it is less efficient at low light levels. With a CCD, though, you can continue to build up the image for more or less as long as you want. Second, CCDs are about 25 times more sensitive to light than a photographic emulsion. This serves to reduce exposure times rather than to combat light pollution in any way. Nevertheless, the ability to record faint stars and nebulosity in an exposure of only a few seconds is a great boon to those who are accustomed to spending long periods glued to the eyepiece, trying to guide the telescope on a star while starlight slowly accumulates on the emulsion. Many a long hour of such effort has been wasted because of poor focusing, a momentary jog of the telescope, or some other problem that did not become evident until the film was processed. And because CCD images can be viewed immediately after they have been recorded, any problems are immediately obvious and steps can be taken to correct them.

A further advantage of CCDs is that they have a linear response to light. As shown in Figure 6.9, the characteristic curve of photographic emulsion is at its least contrasty when you most need it, just above the sky fog. Unlike photographic emulsion, a CCD produces an image which has the same contrast across the whole brightness range of its image. This means that you can subtract the sky background completely. Any

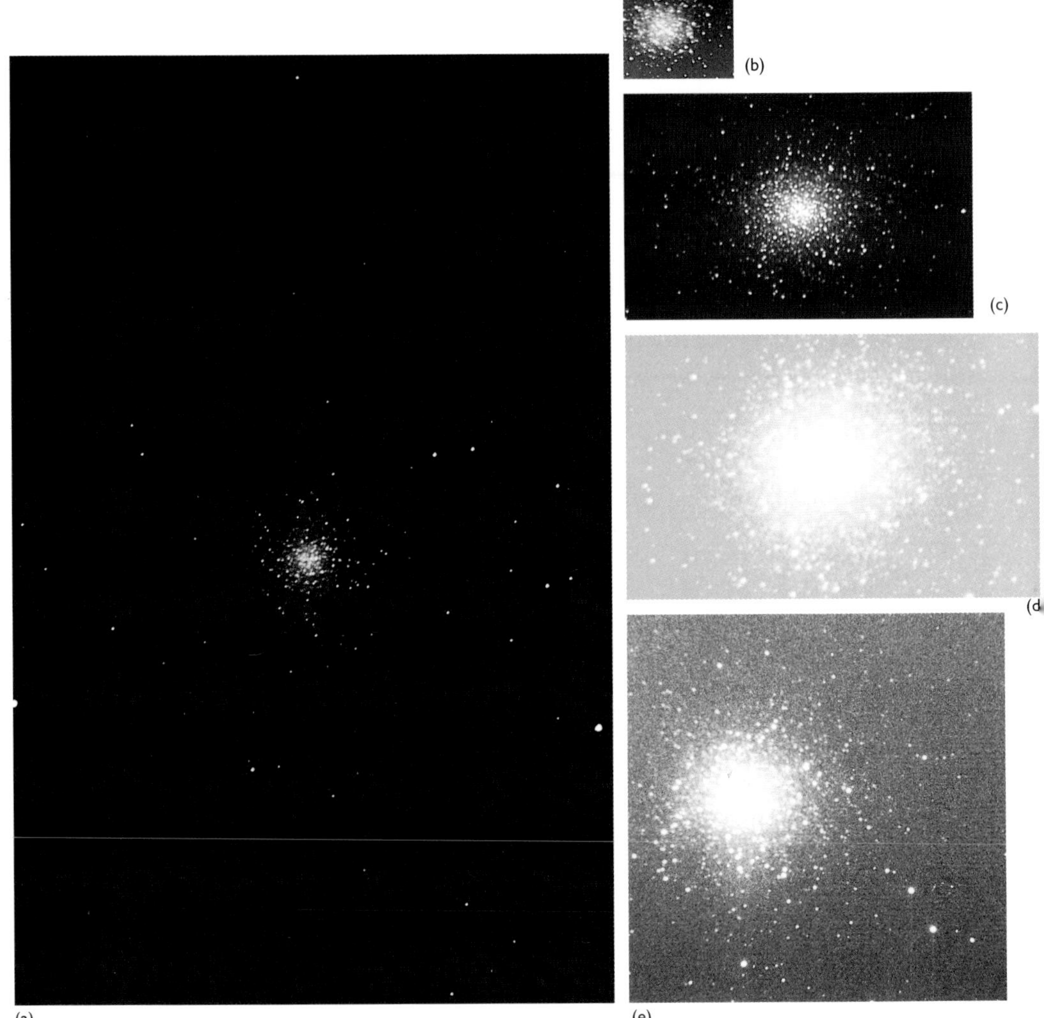

Figure 6.18 *Five images of M13 shown on the same scale, as a comparison of different CCD systems. (a) Direct photograph by Steve Smith from Birmingham, UK, taken on Fujichrome 400 using a Celestron 200-mm (8-inch) SCT at f/10, exposure time 10 minutes. (b) Image from an Electrim CCD (which uses the same chip as the ST-4) on a 250-mm Meade SCT by Nik Szymanek from West Horndon, Essex, UK, 30 seconds. (c) Nik also took this ST-6 image with the same instrument, giving three 120-second exposures in red, green, and blue light and combining them. (d) Starlight Xpress CCD image by Terry Platt from Binfield, Berkshire, UK, with a 200-mm f/5, 160 seconds; darker skies than for the other images here. (e) Phil Hodgins of High Wycombe, UK, used a Hale Research CCD camera on a 250-mm Meade LX200 f/10, 40 seconds.*

recorded image which is even marginally brighter than the background can be made easily visible. It is this ability that makes CCDs so useful to city astronomers.

Finally, because the image is stored digitally, once it is captured you can manipulate it to your heart's content. The equivalent of darkroom work can be carried out at the computer keyboard, and a number of sophisticated image improvement techniques are available.

The combination of these advantages makes CCD astronomy a very powerful technique. Impressive feats are possible even from light-polluted skies. Maurice Gavin has been observing from Worcester Park, a suburb of London, for many years. His observatory is 20 km (12 miles) from the city centre, which places it well within the suburban sprawl and gives sky conditions no better than those obtainable in most city areas. Like many other observers, visually he can see only the faintest trace of the Milky Way when Cygnus is overhead. Maurice has taken many fine photographs with a 440-mm (17½-inch) reflector, but found that all but the brighter objects were being wiped out by light pollution. Since using an ST-4 CCD, he has taken images of a wide range of objects. From his unpromising location, Maurice can reach 18th-magnitude stars with a recording time of just 90 seconds. Although the sky brightness limits each imaging to about this length, any number of images can be combined by processing to synthesize a recording time as long as required.

With such a powerful technique, a huge range of objects can be recorded. Comets that defeat even country observers without CCDs, remote galaxy clusters, asteroids, and faint wisps of nebulosity can all be picked up easily. The ST-4 chip has strong red sensitivity, which also helps to show H-alpha emission. Figure 6.18 shows the globular cluster M13 in Hercules, as recorded by photography and as imaged by various CCDs.

At the other end of the brightness scale, planetary imaging has also been revolutionized by CCDs. As mentioned above, one limitation of photography is the characteristic curve of the emulsion, with its low-contrast "toe." Good contrast is essential to bring out elusive markings on planets, and with a CCD the contrast of even the faintest features is good, as Figure 6.19 shows. Furthermore, you can see the results almost

Figure 6.19 *CCD images of (a) Mars and (b) Jupiter by Terry Platt using a Starlight Xpress CCD on a 318-mm (12½-inch) tri-Schiefspiegler.*

(a)

(b)

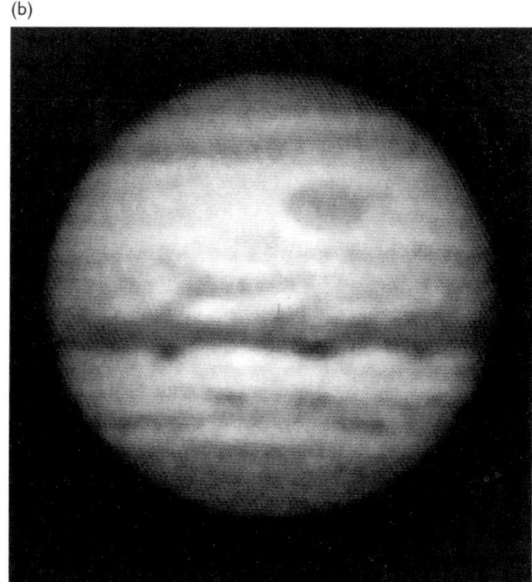

immediately, and save only the best exposures. Exposure times are still of the order of a large fraction of a second, so it is not possible to "freeze" the seeing completely.

Despite all this, the raw image from a CCD can be unimpressive. The digital nature of the image allows a considerable amount of additional processing, with the result that every vestige of information in the picture can be revealed and enhanced. Figure 6.20 shows a basic image processing sequence. False colour is easy to add using software. By using the right palette, you can bring out features in galaxies or nebulae that would otherwise be too subtle to see. Even features one grey level above sky background can be coloured however you choose, either enhancing the image or degrading it, depending on your point of view.

Most CCDs take monochrome images only, though systems which use colour CCDs are available. These are manufactured with tiny colour filters over each pixel, which is the system used in domestic camcorders. Professional broadcast-quality video cameras, incidentally, generally have three separate CCDs, each receiving light of a particular colour from a multi-layer filter assembly. Much the same technique is possible for monochrome CCDs used in astronomy, though colour imaging is usually done by taking three separate exposures through red, green, and blue filters. These are combined using image-processing software, and displayed on a colour monitor. The only disadvantage with this system is that any

Figure 6.20 *Image processing of CCD pictures can be carried out at leisure. Here are three stages in the processing of an image of M81 taken by Terry Platt, as displayed on the computer monitor. (a) The raw image from a Starlight Xpress CCD attached to a 200-mm (8-inch) f/5 Newtonian, exposure time 320 seconds. The output amplifier in the unit head has introduced a spurious glow in the top left-hand corner. (b) The same image after a dark field (no exposure to light) has been subtracted electronically, thus removing the amplifier glow. (c) The result after logarithmic rescaling and normalizing. This compensates for the inability of the monitor screen to display the entire brightness range in the raw image, and emphasizes the contrast and brightness of the spiral arms while retaining detail in the nucleus.*

(a)

(b)

(c)

parts of the image that change (such as the shadow of one of Jupiter's satellites) between exposures will have coloured fringes, unless the three images can be taken in very quick succession.

Nothing is ever perfect, and there are drawbacks to CCDs. The limited number of pixels is probably the most obvious. With only a couple of hundred elements on a side, the picture quality looks coarse compared with a photographic image. This is because the CCD itself is also small, typically giving a field of view less than 10 arc minutes at prime focus on a telescope of focal length 1000 mm. A 35-mm frame at the same focus would cover over a square degree of sky. One solution to this, for extended objects, is to take several images in a grid pattern, and stitch them together to cover a larger area using software. This is not quite as easy as it sounds – each exposure must be identical, the CCD must be "flat-fielded" (exposed to an area of uniform brightness) to overcome sensitivity variations, and the field centres of the images must be arranged precisely across the object being imaged. These problems can all be overcome, but the procedure is more involved than simply making an image.

Another solution would be to use a larger CCD. These are, however, expensive to produce. As manufacturing technology improves, prices for high-quality chips will fall, so in the not-too-distant future a CCD which gives broadcast-quality images may become available for amateur use. The problem then will be storing the image data. As it is, each image from an ST-4 CCD occupies at least 31 kilobytes of data storage. It is possible to store about 45 of these images on a standard high-density floppy disk. A broadcast TV picture, however, requires well over a megabyte of storage space, so only one will fit on a floppy disk. Larger images will require different methods of image storage. It is pointless to speculate on how the power of computers and data storage capacity will change over the coming decade, but we can certainly look forward to greater things. However, it is hard to imagine a CCD-based system managing to reproduce the sort of wide-field image that it is quite easy to take by straightforward photography. Each system has its own merits, and for many purposes CCDs will not replace photography in the foreseeable future.

The small image area of the CCD also means that it can be hard to find the target object in the first place. Visually, from the city all you may see in the finder is a blank piece of sky. Whereas the country observer may be able to identify a few stars and pinpoint the location, from the city you have to point the telescope in approximately the right direction, make an exposure, then hope that you can identify the starfield revealed. This means having access to a fairly detailed star chart. You can now obtain computer-displayed star charts based on the *Guide Star Catalog* (GSC), produced for use with the Hubble Space Telescope and widely available on CD-ROM. These show stars down to 15th magnitude, which is still not as faint as moderate amateur telescopes are now able to penetrate. Again, this demands a yet greater degree of sophistication in your set-up.

The short exposure times possible with CCDs might suggest that the drive rate of your telescope need not be very accurate, but in practice the tiny image area shows up any inadequacies and the drive needs good error correction. In addition, any vibration caused by wind or heavy traffic will blur the images.

Are CCDs cheating? For sheer visual spectacle, they cannot compete with the view of even a light-polluted Double Cluster through a cheap pair of binoculars. Just to see the faint smudge of the Andromeda Galaxy with the naked eye and realize that it is a galaxy over 2 million light years away whose light left it before modern humans walked the Earth, is a thrill even for experienced observers. But astronomy adapts with the tools available, and to be able to track down remote galaxies, see them on your computer monitor, and turn them into pretty red and green blobs with a few keystrokes, is a different kind of thrill.

Photoelectric photometry

If you are keen on variable stars, you may like to enter the more precise world of photoelectric photometry. Professional astronomers have been doing this for years, using standard photomultipliers and filters, and the amateur who wants results that will stand comparison needs to use similar techniques. Like visual variable-star work, it can be carried out just as easily from a city location as from the country.

A number of devices suitable for amateur use are now available, typically costing as much as one of the larger SCTs, though there are cheaper units which use photodiodes rather than photomultipliers. Ideally, the system should use a photomultiplier tube, shielded from magnetic fields and well engineered to prevent it moving as the telescope is shifted from one part of the sky to another. These days, an interface with a computer is advisable. Although CCDs are capable of carrying out photometry, and software for the purpose is supplied with them, their small field of view can limit the number of suitable comparison stars. The high red sensitivity of most CCDs means that unfiltered images are not comparable with visual observations, and the uneven sensitivity across the field of a CCD must be taken into account at each session by flat-fielding the device to a field of uniform brightness and of the same colour as the object to be recorded. This field is then electronically subtracted from subsequent observations. But if these difficulties can be overcome, CCDs can give a photometric accuracy of about 0.1 magnitude.

The telescope can be as small as a basic 200-mm (8-inch) SCT, but the drive must be free from periodic errors: a star should be kept to within 3 arc seconds of the centre of the telescope's field of view during the period of observation, which is of the order of 30–60 seconds. Using a telescope with an aperture of 250–350 mm (10–14 inches), you can monitor stars down to 10th magnitude even from a light-polluted environment.

If the photometer has an accuracy of 0.01 magnitude, a vast range of stars can be monitored and it is possible to take part in serious programmes of work. This sort of thing is a far cry from the casual stargazing which attracts many to astronomy in the first place, and it demands greater rigour and attention to detail than other forms of observing, but it has its own rewards.

Astrometry

The measurement of the positions of astronomical bodies, astrometry, is fundamental to astronomy. It was the main reason for the foundation of the earliest national observatories, and it continues to be valuable today. Although precise astrometry of stars is now carried out by automated equipment beyond the scope of amateurs, the astrometry of Solar System objects is a different matter. Positional measurements of newly discovered comets and asteroids in particular are in short supply. While faint comets may not be observable from city skies even with CCDs, many are easily within range, as are asteroids.

Some dedicated amateurs carry out astrometry by the classical method of taking photographs, then measuring the positions of images using plate measuring machines. Their results are comparable to professional measurements, but painstaking care is needed at all stages and the output rate is low.

Again, CCDs have transformed the scene. Although a star's image may be spread over several pixels, it is possible to measure the centre of the image to a greater accuracy than the size of an individual pixel. There is software for this purpose, doing away with the need for an expensive and hard-to-come-by plate measuring machine. With reference stars taken from the GSC, astrometry to within 0.5 arc second is possible with care. A test carried out by members of the BAA showed that the best results possible with a CCD are comparable to those obtained by traditional photographic methods. The GSC is not a perfect source of reference stars, since it does not include the proper motions of stars (their actual motions through space), so it will get progressively out of date.

CHAPTER 7

If all else fails ...

After all the tips and advice given in this book so far, there is one surefire way to overcome the light pollution and see all those objects you have read about: go where the skies are really dark. Many amateur astronomers now choose this option, and seldom if ever observe from their home. Others divide their time, observing the most accessible objects from home and waiting until they can visit dark skies to look for nebulae and other deep-sky objects.

While it might seem a simple option, travelling to a dark site is not always easy. Some people are lucky enough to live where light pollution is limited to the city area itself, and can get away to dark skies fairly quickly. I once enjoyed a good night's observing just 16 km (10 miles) from downtown Wagga Wagga, in Australia. The town, which is brightly lit, was just a glow on the horizon and everything else around was dark. Maybe the locals would consider that poor conditions, and would go farther out, but it is all a question of what you are used to.

Most people will have to do some travelling even to get to tolerably dark skies, in which the Milky Way is visible and there is at least a fighting chance of seeing faint objects. This probably means a drive of an hour or more, and using the car as a base – the sort of trip you can do in a single night. But for really dark conditions, a major journey is called for, either a very long drive, with overnight stops, or a plane journey. Indeed, for non-drivers every trip to a dark site takes on the flavour of an expedition. In this chapter I look at what sorts of site you might try, and discuss the equipment you need for portable astronomy.

Choosing a site

Satellite views of the Earth at night show us what we are up against (see Figure 1.2 on page 8). The built-up areas of populous countries are marked by sprawling blobs of light. Satellite photographs of Britain show my own location on the fringe of London to be well within a zone of brilliance some 105 km (65 miles) long, centred on London itself. Despite the fact that there is open countryside not far away, this has little effect on the brightness of my skies as there are numerous major roads and small towns in every direction. If I drive 40 km (25 miles) or so towards Oxford I reach the Chiltern Hills, where there is some respite from the light. Then, as I approach Oxford itself, there is another lozenge of brilliance. Even the darkest spots on the Chilterns are hardly good sites; the naked-eye limiting magnitude might reach 5.5 on a good night. But to me, that's pretty dark. I cannot better this unless I travel for about three hours, to the West Country or to Wales. With the British climate, the chances of it still being clear when I get there are slim. My situation is not unique, or limited to Britain – observers in other countries have just the same problem.

The satellite photographs do help in choosing a dark site, but they do not tell the whole story. They show only areas where there are upward-shining lights. The bright patches have an effect on the sky glow well beyond the area they cover in photographs. The "light bubble" from a town extends an enormous distance in all directions. Even in the early 1970s, the astronomers at Pic du Midi could detect photoelectrically the sky glow from Barcelona, 360 km (150 miles) away and on the other side of the Pyrenees. I would estimate that for most purposes you need to be about 80 km (50 miles) away from a major city for the effect of its lights to be negligible for observational purposes. Although you will still be able to see a glow in the sky, it should not be objectionable. On mainland Britain, the only places where you can be this far from all major cities are the heart of Wales and the north of Scotland. To be completely free from the effects of upward-shining lights, you would have to be about 275 km (170 miles) away from them. To get this figure I have made the arbitrary assumption that the densest part of the atmosphere is within 5 km (3 miles) of the surface; almost exactly half the atmosphere is below 5 km. An air molecule 5 km above you will have a horizon 275 km away, and so will just be illuminated by the lights, however faintly, but the air mass below it will be hidden from the lights by the shadow cast by the Earth.

The coast might seem a good place to head for to get away from light pollution, but a photograph I took at Portland Bill (see Figure 7.1), a promontory on England's south coast, shows that this is not the case. The

Figure 7.1 *Light pollution extends far from the actual lights, as is shown by this photograph of the nighttime view south from Portland Bill. Although the site is on a promontory that juts out several miles from the south coast of England, sensitive film records star trails in a pale blue sky tinged with yellow, and a band of cloud along the horizon. The bright line along the horizon is the trail of a ship's lights.*

French coast is 100 km (60 miles) away, yet the time exposure of the sky to the south shows the unmistakable sign of light pollution. The lights of the English south coast resorts shine out a considerable distance.

The choice of observing site depends not just on the darkness of the skies. If you are looking for a site within reasonable driving time from your home, and plan to return home after the night's observing session, you will not want to travel for much more than an hour or two unless you can sleep in your vehicle. You will also need to find a site well away from other people, for various reasons. You do not want car headlights shining on you just as you have become dark adapted, and furthermore, although it can be fun to show the wonders of the heavens to inquisitive strangers, they can interfere with your programme of work. Some American amateurs have even started to carry firearms as a precaution against potentially dangerous situations. To those in other countries this is unthinkable at present, but one wonders if it will become the norm everywhere in years to come. I have encountered people with guns at night in Britain, but they were interested only in rabbits. I always hope they can tell the difference between a rabbit and an astronomer.

It was to avoid such problems that some years ago a group of amateurs in the English Midlands devised a sign, in conjunction with the local police force, which they could place near their observing site to warn people that there were astronomers about. This followed an incident in which one of them was pounced upon by an alert constable attracted by the strange nocturnal activity! Thankfully, British police do not as a rule carry guns, or the outcome might have been more serious. Several times I have been the centre of police attention while observing in the country, but have found that once they know what I am doing the only problem is getting rid of them. The inhabitants of country areas can be inquisitive, and are often concerned about what might be going on in their area, even if they do not own the land. The obvious answer to these problems is to observe from private land with the agreement of the owner. Although this may take some organization, it could make for a much more pleasant and comfortable session, especially if you can observe in the lee of outbuildings on windy nights. Some astronomers in Britain lucky enough to have good sites themselves advertise their facilities to others, and offer bed-and-breakfast accommodation along with the use of telescopes. Look in the classified ads of astronomy magazines for details.

The search for dark skies

Every so often, the desire for really dark skies overwhelms us. For amateurs living in the conurbations of Britain or northern Europe there are many possible sites, but it is noticeable from light-pollution photographs taken from orbit that France is considerably darker than most of her northern neighbours. It is just that there are fewer streetlights, though main routes into towns are as well lit as elsewhere in Western Europe. The main difference seems to be in the level of suburban lighting and in villages. For British observers, the simple act of crossing the Channel brings darker skies, once you are away from the towns. And the farther south you go, the better the chance of good weather.

There are now several observatories for the use of amateur astronomers, of which the best known is that at Puimichel, about 70 km (45 miles) northeast of Aix-en-Provence (see Figure 7.2). It offers basic accommodation and the use of several instruments up to 1 metre (40 inches) aperture, though the 1-metre itself is in great demand during moonless summer nights. It is best to go in a group, which gives you more opportunity to use the main instrument than if you went alone. You can set up your own telescope in an area reserved for the purpose. Although the skies are not perfect, the facilities and the welcome extended to all amateur observers make the site very popular. The best skies are in the winter months when the mistral blows – a cold air current streaming down off the Alps and into the Rhône valley, bringing crystal-clear skies. The observation area is provided with large wooden windbreaks to make life more comfortable in these otherwise excellent observing conditions. Puimichel is not far from a main autoroute linking it with northern Europe. It is possible to drive there in a day, even from southern England.

Another observatory for amateur use is in Portugal's Algarve, on the Atlantic coastline. Known as COAA, the Centre for Observational Astronomy in the Algarve, it is run by a British couple. It has fewer instruments than Puimichel, and can cater for up to ten people at a time, though the accommodation is more luxurious than at the French site. The main instruments are two 300-mm (12-inch) reflectors, one equipped with an ST-4 CCD, and again people are encouraged to bring their own instruments. The Algarve has one of the best climates in Europe, boasting 3000 hours of sunshine each year (three-quarters of all daylight hours). It is a major holiday area, and there is a little light pollution from the nearby resort of Portimão. It is a long journey by car from the built-up areas of northern Europe, and most people visiting go by plane, taking a cheap charter flight. An example of the deep-sky work possible from COAA is shown in Figure 7.3.

Figure 7.2 *The French hilltop village of Puimichel is host to large numbers of amateur astronomers from all over Europe. The dome of the 1-metre (40-inch) reflector can be seen to the right of the village.*

Figure 7.3 *If your taste is for photographs of galaxies, then you must visit a dark site. This picture of the galaxies M65, M66, and NGC 3528 in Leo was taken by Howard Brown-Greaves through the 300-mm (12-inch) f/5 reflector at COAA in Portugal. Exposure time 22 minutes on ScotchChrome 800.*

Another site popular with European astronomers, both professional and amateur, is the Canary Islands. These Spanish-owned islands off the west coast of Africa are a four-hour flight south, at latitude 28°, so they offer much more of the sky than can be seen from northern Europe. The main island of Tenerife has the advantage of being accessible by charter flight, with package deals on accommodation which can be much cheaper than trying to book your own, particularly as transport to the resorts is included. The coastal strip is ablaze with lights (see Figure 7.4), so do not expect to be able to observe much from your hotel balcony. However, Tenerife rises steeply from the sea, and just an hour's drive from the main resorts will take you to an altitude of about 2000 metres (7000 feet), often above an inversion layer kept low by the cold sea current. There is plenty of room to observe from above the treeline, in Las Cañadas National Park, and even in winter the temperature at this level rarely drops below 5°C (41°F).

Despite the presence of so much lighting, the skies from the centre of the island are often excellent by European standards (see Figure 7.5), and a large professional observatory has been built on the ridge. The main peak of Tenerife is the volcanic Pico del Teide, at 3718 metres (12,198 feet), but there is no point in trying to observe from its summit. When British astronomers were site testing for a Northern Hemisphere Observatory, as it was then known, an intrepid team camped near the top of Teide to see what it was like. Among the hazards were the sulphurous fumes which filled their tent. A less uncomfortable site was

Figure 7.4 *Tenerife is favoured by many European astronomers as a relatively dark and accessible site. This view from neighbouring La Palma shows how the coastal fringes are strongly illuminated, though the central caldera area of Las Cañadas can still have dark skies under the right conditions.*

the nearby peak of Guajara, which was chosen by Charles Piazzi Smyth for his observations of the 1834 return of Halley's Comet, and from where he also made the first infrared observations of the sky. The remains of his site can still be seen. But there is little advantage to be gained from climbing there since Las Cañadas offers many more accessible locations. You can even stay in the Parador Hotel below Guajara. Although this is expensive by Canarian standards, it does avoid having to make the long and potentially dangerous drive down to your resort after a night's observing.

The major Canarian observatory is on the island of La Palma, a short flight from Tenerife. The observing conditions on La Palma are very good, but as from anywhere in the Canaries there is always a strong possibility that visibility will be impaired by Saharan dust suspended in the atmosphere. The dust seriously affects observation on about one night in every ten, and is detectable on virtually half the nights in the year. Groups who wish to observe from the site, which occupies most of the summit area from which observations are possible, should contact the secretary of the Observatorio del Roque de los Muchachos well in advance. As with all working observatories, the staff there have their own programmes and the telescope time is very expensive, so it is important to go through the correct channels. Daytime visits to the telescopes are possible by prior arrangement, but you must arrange your own travel to the site. The road can be very dangerous in winter, and occasionally the observatory becomes inaccessible, even by four-wheel drive vehicle.

Figure 7.5 *The Milky Way is a splendid sight from a dark site. This photograph was taken by the author from near Teide observatory on Tenerife using a 24-mm lens at f/2.8, exposing for 6 minutes on Fujichrome 1600 film. The horizon slopes because of the angle of the equatorially mounted camera. The white lines near the horizon are the lights of aircraft from the airport on the nearby island of Gran Canaria.*

There are fewer such facilities available to North American amateurs. This is probably because, unlike Europeans, American observers can generally find dark skies within an acceptable driving time of their home. There is the Star Hill Inn near Sapello, New Mexico, which has six cabins, very dark skies, an astronomy library, and telescopes to rent, including a 440-mm (17½-inch) Dobsonian and a Celestron C14. However, a popular feature of amateur astronomy in the US is the star party – one-off or regular observing sessions held at dark sites, generally reached by camper van. There may be up to 2000 amateurs at such gatherings, which offer a daytime schedule of lectures and other events.

As well as the opportunity to observe from dark skies, star parties offer the chance to compare commercial and home-built equipment. Some are attended by representatives of the major telescope firms, eager to give their goods a high profile, with complete telescope systems occasionally given away as prizes. And other amateurs who already have the equipment will gladly give their verdicts on performance. In many cases you will be able to compare for yourself, with the owner's permission. There will undoubtedly be an impressive array of telescopic firepower available, and some proud owners are only too pleased to see a long line of prospective viewers, waiting for a brief peek at M57 or whatever happens to be on view. The biggest instruments are the most popular, not surprisingly, and those humble souls with mere 200-mm (8-inch) commercial reflectors will probably hide away in shame.

There is an etiquette to star parties, chief among which is not to use any white lights while observing is in progress. In particular, this means not driving your car on site at night, so make sure you arrive in good time, and do not expect to be able to drive away when you get tired. Many star parties are intended for serious viewing, so if you have a family make allowances for their needs, unless they are all as dedicated as you. (Other aspects of star-party etiquette are given in *Sky & Telescope*, September 1993, page 90.)

Among the big regular star parties and telescope conventions are the Riverside Telescope Makers' Conference in California, the Texas Star Party (both in May), Stellafane in Vermont, intended mainly for serious telescope makers, Mount Kobau Star Party, British Columbia (both in August), and the Winter Star Party in Florida (January or February). In addition, the Royal Astronomical Society of Canada organizes a star party on a different site each year. *Sky & Telescope* lists events each month; contact addresses are given at the end of this book. You may need to book your place well in advance, since some of the big parties have to limit attendance.

As well as established sites such as these, international trips are often arranged to dark-sky sites such as the Australian outback or for events such as eclipses. These are advertised well in advance in astronomy magazines, and offer you the chance to get to dark skies and to meet other astronomy fanatics. The downside of these trips is that they can remove your personal freedom to choose your own site.

I have often wondered where the best skies in the world are to be found. In my own limited experience, the southern hemisphere has better skies than the northern. This is certainly the initial impression you get, partly because the Milky Way in the southern hemisphere is brighter than the northern section, and partly because your attention is drawn to familiar constellations higher in the sky. But when comparing the skies from, say, Tenerife and Brisbane, which are at similar latitudes north and south of the equator, the impression persists. I suspect that the lesser amount of industry and, in particular, air travel has a lot to do with it. Every aircraft produces a trail of pollutants at high altitude, which provides nuclei for condensation and hence leads to milky skies.

Professional astronomers are in the position of being able to travel to the world's best astronomical sites. Some say that the mountain sites of Chile, where there are several European and American telescopes, are the best they have experienced, while others maintain that the skies at the South African Astronomical Observatory at Sutherland, on the Great Karoo plateau, are better still.

It is now becoming increasingly possible for professionals to observe remotely. They can sit in their office in, say, Edinburgh on a rainy day and conduct observations being carried out at nighttime atop Mauna Kea in Hawaii. A data link enables them to see what is going on and assess the quality of the observations. Such remote observing is now being extended to amateurs, using large aperture instruments at prime sites such as Mount Hopkins, Arizona, and Mount Wilson, California. This can be thought of as an extension to CCD and computer-controlled astronomy, where the observer sits watching the image appear on a computer screen rather than at the eyepiece. Instead of the telescope being a metre or so away, it can be a continent away. Naturally this sort of facility costs money, but some city planetariums and science centres are developing the idea and may provide the funds. Anyone undertaking a serious programme of work which would benefit from such an arrangement could find themselves at the virtual controls of a telescope which not long ago was the exclusive preserve of professionals.

Your car – a mobile observatory

Travelling to your observing site by car has the advantage that you can take all the gear you are likely to need for the night's observing. A car is very convenient for this, but it is hardly purpose-built. There are a few modifications you can make before setting out for the rural darkness. First, get a roll of translucent red sticky tape, and use it to cover your car's interior lights and any other lights you use during the session.

Next, pay attention to your power supplies, for both your telescope drive and any camera you are going to use. Telescope drives draw very little current from a car battery, so in theory it is unlikely that you will drain it. However, I have been caught out on a remote road in New South Wales when the battery of my hire car packed up after half an hour on parking lights. The engine would not turn over at all. Fortunately, after an hour the battery miraculously recovered, but I have never completely trusted a battery since.

It is a good idea to use a separate battery for the telescope drive – it need not be a new one. Just keep the old battery when you get a replacement. But bear in mind that it should be fixed in place in some convenient location, and make absolutely sure that nothing can drop across the terminals and short them out. Using a separate battery means that you can do away with that maddening cigarette lighter connection. It is strange that just about the only universal fitment on a car is also the most infuriating. Although it may do the job of lighting cigarettes well enough, the problems start when you try to use it as a general connection for all the accessories that are now available. It is not permanent enough for my liking, and the plug can easily be pulled out. Furthermore, in addition to the telescope drive you might wish to use a number of accessories that require electric power: an extra light, a portable hair dryer to get rid of dew on the camera lenses, a computer, maybe even a soldering iron to fix the connections that always come undone as you observe. You may also find that the socket works only when the ignition is switched on. But do not leave the ignition on for long periods if the engine is not running, as this continuously energizes the ignition coil, which could overheat and burn out. Unfortunately there are few alternatives.

It would be best to have a different system for low-voltage power supplies, with a bank of sockets conveniently located at the back of your car so that you can simply plug in whatever you want. I do not know of any particularly good connection system. Power connections for small accessories these days are often via little jackplugs, but they are just as bad in their own way. Apart from the fact that they are very difficult to wire up, for some perplexing reason manufacturers have reached no agreement on whether the central pin should be positive or negative. The popular Super Polaris mounts have a negative central pin, the opposite of some other appliances which have identical plugs. The temptation to use ordinary household plugs for low voltage should be resisted, as there is then the danger that someone – maybe even yourself – will plug your low-voltage appliance into the household supply at some stage. It looks as if we are stuck with the cigarette lighter system, with all its faults. It is a good idea to make up a socket bank wired directly into either the car battery or a separate battery, with a device for preventing the plugs from being pulled out. One simple method is to cut a narrow slot in a piece of wood fixed to the bank of sockets. Tie knots in the cables and slip them into the slot so that if the cable is pulled, the knot prevents the plug from being pulled out.

For safety's sake it is best to stick to low-voltage devices. But should you need a higher voltage to run your telescope or even your computer, you can do so via a device called an inverter. These are sold for use on caravans and yachts, and provide fairly low power at domestic supply voltage from a car battery. They are similar to the variable-frequency oscillators used to control synchronous motor drives, but they operate only at domestic supply frequency. Bear in mind that the more powerful the unit, the greater the temptation to plug in more and more appliances, and the greater the risk of draining the battery.

In addition to your observing equipment, it is a very good idea to take a small portable table, such as a card table. This will take your charts, observing book, pens, and of course eyepieces, filters, and so on. You may like to rig up a canopy over it to protect everything from dew or frost, together with a small red light. Improvements such as these make a great difference to the ease and pleasure of observing out in the field. It is also essential to have a white light, either a torch (flashlight) or one run from the car battery on a long cable, to help find those vital accessories that always seem to get dropped. However, if you are going to be observing with other astronomers around, be careful when using it or you could make yourself very unpopular. A good magnet is well worth having, to help retrieve screws and bolts that you drop when fumbling with cold hands to fix a telescope to its mount. It is worth painting the heads of screws and bolts red or yellow. A black or even a shiny object has a knack of disappearing in the grass when it falls to the ground.

I shall gloss over some of the other paraphernalia of observing in the country, such as warm clothing, insect repellent, refreshments, and so on. There is also the matter of accurate polar alignment, which is particularly important if you want to attempt photography. Many observing guides tell you how to do this; all I shall add is that if you have a mounting with a polar alignment system, take along a piece of old carpet which you can lie on. This will also come in useful when you are trying to line up your camera on a particular area for piggyback photography.

It sounds obvious to say "make a list of everything you will need," and I am the first to ignore this advice. But if you have no list, there is one item you should always double check if you are using a German mounting – the counterbalance. These are probably the most easily forgotten items, followed by the removable shaft on which the counterbalance slides.

Lastly, do not forget that if you are hoping for good seeing, there is no point in observing in a line of sight that passes over the car, for its cooling engine will continue to give off heat long after it has been turned off. More obviously, avoid a line of sight that passes near the exhaust if the engine is running. And if you leave your camera on a driven platform behind the car, carrying out a time exposure while you sit inside with the engine running to keep yourself warm, you may well find that the camera is immersed in a cloud of condensation from the exhaust.

The portable telescope

Whether you travel to your dark site on foot, by car, or by plane, you may well be taking a telescope with you. Any telescope which is not permanently fixed on a massive mounting is a compromise in one way or another, and you have to decide for yourself what suits your own circumstances. There are various possibilities.

To my mind, a telescope is portable only if you can easily pick it up, mounting and all, with one hand. Everything else is merely *trans*portable. Horace Dall was the expert at truly portable telescopes. His finest was a 150-mm (6-inch) reflector which he could conveniently carry in a raincoat pocket. This is no exaggeration. He travelled all over the world with

Figure 7.6 *The author's portable 250-mm (10-inch) Dobsonian telescope. A single aluminium spar carries the secondary and eyepiece. Focusing is carried out by moving them up and down the spar. This form of focuser is easy to make, as it does not require accurate machining.*

it, and it was never noticed by customs officials, though that was in the days before metal detectors at airports. The design was a "thin-mirror Dall–Kirkham." The secondary was carried on an arm which moved along a short square-section beam for focusing. It fitted easily to a small photographic tripod. To make such a device commercially would be prohibitively expensive, and most of us have to settle for more mundane instruments.

Undoubtedly the most popular is the 200-mm (8-inch) SCT. The basic system fits in a carrying-case, along with its wedge (which converts its basically altazimuth mount into an equatorial). There is also the tripod. For a basic Celestron C8 Plus, which was designed for portability, the total weight comes to 17 kg (38 lb). This easily stowed in even a small car, which accounts for the popularity of the system. A more expensive equatorial head adds little to the overall weight. Surprisingly, the smaller Celestron C5 is only a few pounds lighter. The standard Meade 2080, on a heavier-duty tripod, weighs in at 24 kg (53 lb).

Medium-sized refractors – ones of, say, 100 mm (4 inches) aperture – can also be quite easy to move around. The tubes are often quite short, and the rest of the apparatus is similar to that of the SCT.

People often buy one or the other of these two because they need to travel by car to a dark site and want something easy to carry. If they want larger apertures, they start thinking in terms of Dobsonians. This could be a mistake. The Dobsonian design has several advantages, the main one being that it is cheap to make. However, it is not necessarily intrinsically lightweight, for the rocker box and cradles are usually quite heavy, unless they are specifically designed for lightness. Think carefully about what you want. You may find that a standard Newtonian of similar aperture is just as easy to carry, if not easier. In fact the decision that the telescope has to be able to fit into the car necessitates some compromise on aperture, but perhaps some people compromise too much. The main restriction is on the size and weight of the main tube.

If your chief interest is seeking out deep-sky objects from a truly dark site, you do not need as massive a telescope as the standard Dobsonian. Figure 7.6 shows the 250-mm (10-inch) Dobsonian I made for carrying in the car. It takes up very little room. All the weight of the telescope is in the rocker box and cradles, which are as low and squat as possible. This is because there is no tube or frame as such, just an aluminium pole on which the eyepiece–flat assembly moves up and down for focusing. The secondary assembly is carried on a single strut rather than a three-vaned spider, and there is a light shield made from an aluminium printing plate, fixed to the far side of the secondary mount. It is in the light path, but as it is very thin and edge-on to the light path, it is not much of an obstruction and diffraction effects are minimal.

This design is hopeless at high powers – the aluminium pole vibrates too much, and you cannot get very good alignment – but for powers under ×100 it is fine. The main thing is to save as much weight as possible at the top end, which means using a lightweight finder and avoiding heavy eyepieces. The advantages are that as the whole telescope occupies very little space you can carry it with you on the off-chance that you might have the opportunity to observe, and it takes only a few minutes to set up. More elaborate designs use a Serrurier truss system to hold a top end-ring. In this system, widely used on large professional telescopes, eight struts arranged in triangles provide a rigid support for the top assembly, which carries the eyepiece and flat.

Of course, if you have a large enough vehicle, and can enlist help in setting up, you can transport very large instruments. At North American star parties, 500-mm (20-inch) instruments are not unusual these days. Californian amateur Steven Overholt, for example, has designed and built a type of Dobsonian he calls a "lightweight giant." His aims of lightness and portability were admirably displayed in a 560-mm (22-inch) telescope that weighs just 40 kg (89 lb). It can be dismantled and loaded into a small hatchback, and re-erected in a matter of minutes.

But most large instruments are a lot heavier and a good deal more cumbersome, and most people would not want to carry them around regularly. The most useful telescope is the one you can observe with easily, as William Herschel found when he built his giant 48-inch (1.2-metre) reflector, then the largest in the world. It was so unwieldy in use that he preferred his smaller instruments most of the time.

Taking instruments by air

Travelling by air is by far the quickest way, and often the cheapest way, of getting to dark skies. On many international flights there is a baggage limit of 20 kg (44 lb), which does not go very far where equipment is concerned. A basic 200-mm (8-inch) SCT will take you to the limit by itself. Furthermore, the hassle of air travel is increased the more you carry: you have more to look after, and there is always the risk that an item will get lost, stolen, or sent to Bangkok.

If you are over the baggage limit you may be charged a considerable excess, which may be based on the first class fare rather than your economy class ticket. More often than not you will get away with it – I have only once been charged, and that was on the inter-island flight from Tenerife to La Palma, with the plane only half full. It looked as if the operator was trying to extract as much revenue from the passengers as possible. The best way is to travel with other passengers who are carrying less than the limit, and make sure that you check in at the same time as one another, because airlines generally allow groups travelling together to pool their weight allowances.

One of the biggest dilemmas when travelling by air is the care of your film. If you want to take a large number of photographs you will probably need to take your own film with you, particularly if you use anything other than standard print or slide film, and bring it back home for processing under controlled conditions. This means that your film will have to pass through X-ray scanners, several times if you have to take connecting flights. The machines in use at most airports are claimed to be safe for film up to something like ISO 1000, which is not reassuring if you use ISO 1600 or 3200. As a test, I once arranged for ISO 400 film to be repeatedly passed through an X-ray scanner, and could find no detrimental effect. Photography magazines run articles every summer quoting people who say they have had film damaged by X-rays. I suspect that most of this damage had other causes, but it is always possible that a particular machine has a fault and can damage sensitive film.

A committee of British professional photographers recently carried out an exhaustive test with modern X-ray machines used by the British Airports Authority. Films of speeds up to ISO 3200 made by Agfa, Fuji, Kodak, and Ilford were passed through up to 32 times. In addition, some films were sent on an intercontinental flight to give them authentic exposure to actual conditions. After the processed films had been measured for density variations, the conclusion was that any effects caused by the X-ray machines were smaller than the natural variations in processing. However, the committee warned that not all X-ray machines can be expected to perform as well as the ones in the test, particularly in poorer countries.

It is not advisable to commit your film to the baggage hold, since there it can be X-rayed without your knowledge by machines much more powerful than those used in the airport buildings. I have not had much success in recent years in trying to get security staff to inspect films by hand, though that would be by far the best solution. Usually, however, I carry only a few very fast films, so these I keep in my pocket and allow the rest to go through the machine. The small amount of metal in the film cassettes has so far not set off the metal detector. Keys and money are the usual culprits, so I hand those over if there is a problem. Even if you have to take the film out of your pocket, it is usually held by someone standing on the far side of the X-ray scanner and is not passed through.

It may be possible to arrange in advance with the airline for your films to be hand checked. This would be worth trying if you have more than a few cassettes of very fast film.

CHAPTER 8

Indoor astronomy

With the best will in the world, the amateur astronomer cannot observe every night. Indeed, for the city-bound observer good opportunities to see the sky are infrequent, leaving plenty of time on cloudy nights, or when the conditions are otherwise unsuitable, for pursuing astronomy from the warmth and comfort of indoors. And there are those whose enthusiasm does not extend to observing sessions on every clear night.

Through the glass darkly

If your reason for not going out is that it is just too cold, well, you can actually do some observing – and perhaps even make discoveries – from indoors. If you are incapacitated and unable to go outside, observing through a window may be your only option. Every textbook on amateur astronomy tells you that it is a big mistake to observe through an open window, because the turbulence created as air escapes from your warm room into the cold night will destroy the seeing. Of course, they are right – but only up to a point. There is usually no mention of observing through a *closed* window.

First, this question of seeing. As pointed out in Chapter 2, seeing is all-important to the planetary observer. If you want good seeing, you must avoid temperature variations in your vicinity. Even if you open the window well before you start to observe, the room will probably still retain a large amount of heat, even in winter. But I daresay most observers have pointed a telescope out of the window at some stage, and found that the seeing is not that bad. There is something to be said for it – a warm and comfortable observer can make much better observations.

In fact, as long as you insulate the interior from the exterior sufficiently there is little wrong with indoor observing. Astronomy author James Muirden once made a coelostat so that he could observe the planets from his south-facing study window. A coelostat is simply a flat mirror which reflects the light from the object under study into a fixed telescope. The mirror is slowly rotated in such a way as to compensate for the apparent rotation of the celestial sphere. In the house where James was living at the time, the pitched roof of a porch was just below his study window, ideally positioned for the coelostat (see Figure 8.1). A small refractor was angled downwards towards the mirror. The eyepiece end was inside the study, and one looked into it as if it were a microscope eyepiece. I have never before or since observed in such comfort.

The main drawbacks with this system are that the light is dimmed somewhat by the mirror, and the image is laterally reversed. Finding planets was fairly easy, but it would be more difficult to locate deep-sky objects. Also, observations are limited to a smaller part of the sky than with a conventional telescope, depending on where the coelostat is sited. Photography through a such a fixed telescope should be a good deal easier, since the camera can be mounted very firmly and there is no risk of vibration. The main expense is the flat mirror – it has to be of high optical quality, which can make it almost as expensive as a similar-sized paraboloidal mirror for a telescope.

If that all seems too much bother, you might care to try observing through the glass of a closed window. Modern float glass can be of quite good optical quality, and while it is not good enough for use as a flat mirror, it can have comparatively small effects on transmitted light. As long as the power is not too high and you are not observing at too shallow an angle to the glass, the image can be surprisingly good. A little contrast and light are lost, but this matters only if you are attempting very difficult observations. With low powers, such as with binoculars, you will notice almost no difference when observing through a good window.

Though your view is restricted, it is still possible to carry out useful work through a window. Variable-star observing using binoculars is one possibility. There are even two examples of a discovery having been made

Figure 8.1 *For indoor comfort while observing, arrange a silvered flat mirror below a south-facing window so that it reflects light into a telescope set such that its eyepiece is at a convenient viewing angle. The drawbacks of this system are that only a comparatively small part of the sky can be observed, and a suitable flat mirror (which should be rather larger than the telescope's objective, to allow for observing at an extreme angle of reflection) of sufficiently high optical quality is likely to be expensive.*

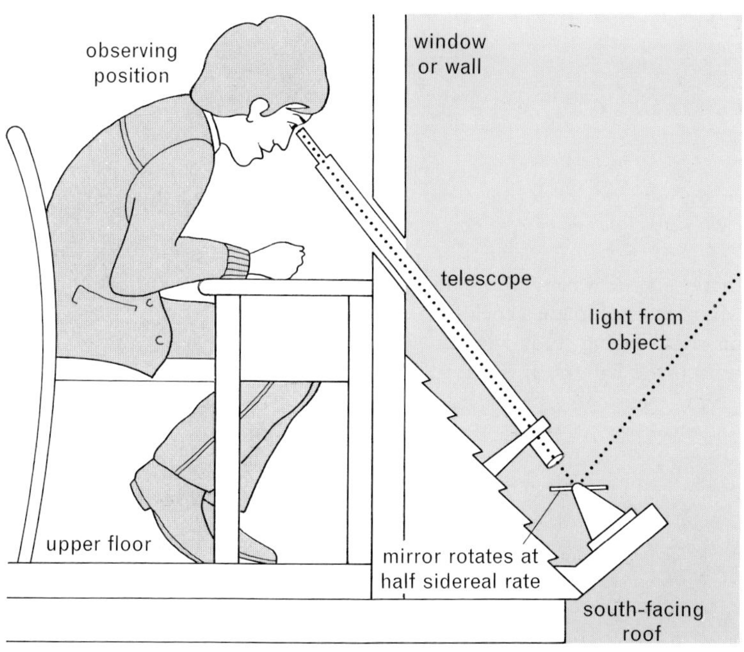

Figure 8.2 *George Alcock's record of novae and comet discoveries is all the more impressive because of his unpromising site near Peterborough, UK. For many years there were high levels of industrial pollution from nearby brickworks, now disused. Now in his eighties, George prefers to carry out his binocular sweeps through double-glazing. Instruments for collecting weather data fill his garden.*

through a window. In 1961, comet expert Michael Candy was testing a telescope by pointing it through a bedroom window when he spotted a fuzzy object that should not have been in the field of view. It turned out to be a new comet, now designated Comet Candy 1961 II. More recently, in 1991 veteran nova- and comet-hunter George Alcock, who – as Figure 8.2 shows – now prefers the comfort of observing from a warm room, discovered his sixth nova, in Hercules. He used 10×50 binoculars through a double-glazed window to spot the fifth-magnitude newcomer! This was hardly a chance discovery – in the previous 27 months he had spent a total of 553 hours searching for just such an object. He spotted the nova at 4.35 a.m., the sky having cleared just half an hour previously.

He also discovered Comet IRAS–Araki–Alcock while observing through double glazing.

It should be possible to carry out driven photography of the stars from indoors, using a telephoto lens of up to 135 mm focal length, though I have not heard of this being done. It would be more difficult to align the mount's polar axis if you cannot see the pole through the pole-finding system, though there are other ways of aligning mounts which are covered in manuals on amateur astronomy.

Observing in other ways

Radio astronomy is a major part of the professional scene, but it has had little impact on amateurs. It lacks the romantic appeal of visual astronomy – there are no dramatic images, just records of radio intensity. There is also the drawback that the feeble radiation at radio wavelengths from celestial bodies usually needs enormous collecting devices for its detection, as evidenced by the giant dishes used by professionals. Nevertheless, there are amateur devotees who achieve good results in a number of different fields.

Meteors can be observed by radio with the simplest of equipment – an ordinary FM radio, preferably one with digital tuning. It also helps to have an additional outdoor aerial or antenna. The technique is simply to choose a radio station which transmits below 91.1 MHz yet is too distant for you to receive directly – maybe one 1300–2000 km (800–1300 miles) away. The frequency must otherwise be clear. A digital tuner helps you to tune to the correct frequency without being able to hear the station. Normally, you will not receive signals from the station, but if a meteor dashes through a region of the upper atmosphere between you and the radio transmitter, the signal will be reflected by the meteor's ionization train, and you will suddenly pick up a short, coherent burst from the station. The burst, known as a "ping," may last only for the briefest instant, but if a meteor leaves a substantial train you could pick up the distant station for several seconds.

It helps to know of a variety of stations in different directions, because the best reflections occur when the meteor is broadside on to both you and the station. This means finding a station at right angles to the direction of the radiant. On a night when a rich meteor shower is near maximum you may detect up to 100 meteors an hour in this way. This technique is useful on those frustrating occasions when a very high hourly rate is predicted, but it is either daylight or cloudy. Then you can keep tabs on the shower's activity from your armchair. (Further details are given in *Sky & Telescope*, August 1992, page 222.)

Dedicated solar observers can monitor the Sun's radio output. Solar observing is a good way to start real amateur radio astronomy. You will need a receiver operating at a frequency at which the Sun is active, such as 136 MHz, and an antenna you can point at the Sun. You can obtain converters which turn the 136 MHz signal into 28 MHz, which is detectable by a standard communications receiver. The antenna is a simple Yagi array – basically, the same as is commonly used for terrestrial TV reception, but with differently spaced elements. With this apparatus you will hear a steady hiss from the receiver, which is not very impressive. To make meaningful observations, you will need to drive the antenna to follow the Sun continuously, and to monitor the signal from the receiver. Until recently this meant using a chart recorder, a device that draws a wiggly line on a chart. These are now being superseded by computers. You may be able to get a second-hand chart recorder cheaply, or if you already have a computer you may prefer to buy or make an interface that will accept the signals from the receiver and turn them into digital form which the computer can store.

This sort of thing is fairly well removed from standard amateur astronomy activities. You will not usually find the apparatus advertised in the astronomy magazines, so unless you are already familiar with electronics and computers you will need to put in some effort. You may be rewarded by advance radio warning of visual solar activity, which you can then tie in with your white-light and hydrogen-alpha observations.

Radio observations, however, are likely to be hindered by nearby sources of electrical interference. Your neighbours' electrical appliances, badly suppressed car ignitions, electric trains, and maybe even central heating controllers could all disturb your records. These disturbances can happen in country areas too, but they will be worse in more densely populated districts.

While it is possible to observe other bodies in space at radio wavelengths, the level of signal from anything except the Milky Way and powerful sources such as the Crab Nebula is too small to be detected without a rather large collecting dish.

There is, however, a very simple means of monitoring a different form of activity, and the basic apparatus is almost laughable – part of it is a jamjar (see Figure 8.3). A bar magnet suspended so that it can swing freely will align itself with the Earth's magnetic field. The streams of solar particles that cause aurorae also produce changes in the Earth's

Figure 8.3 *A jamjar magnetometer is simple to set up and to use. Small variations in the Earth's magnetic field are made visible by the "leverage" provided by the reflection of a light beam from a bulb. To take a reading, you look across the top of the scale and note where the reflection of the bulb strikes it.*

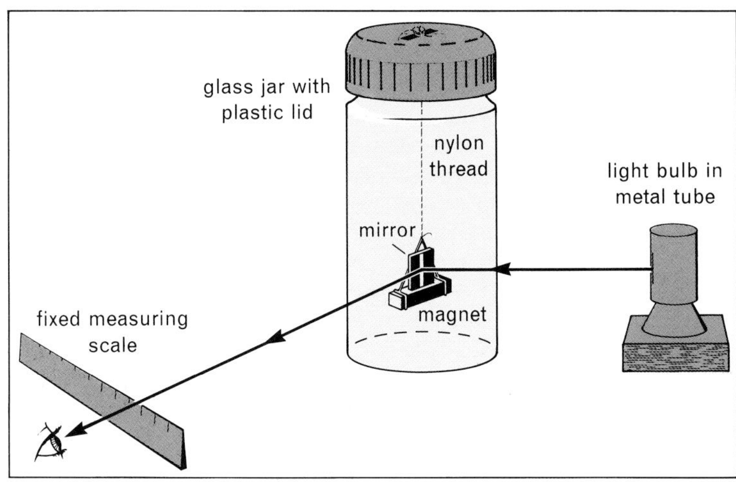

magnetic field which are detectable at geomagnetic latitudes lower than those from which the aurorae are visible. These changes show up as tiny shifts in the magnet's position, which would be hard to detect just by looking at the magnet, but there is a simple method of making them easy to see. Attach a small household mirror to the magnet, shine a light onto it, and watch the reflection of the light beam on a screen some distance away. (Further details are given in *Sky & Telescope*, October 1989, page 426.) During a magnetic storm the beam will shift visibly by several millimetres in a matter of minutes.

A more advanced electronic device, such as a *fluxgate magnetometer*, can pick up the actual commencement of a magnetic event, and can therefore deliver early warning of an aurora more effectively than the simple jamjar magnetometer. Even if it happens to be cloudy where you are, or your skies are not good enough to allow you to see an aurora, by monitoring a jamjar or fluxgate magnetometer you will be able to alert observers who are located where it might be clear. However, both devices depend for their successful operation on being in a reasonably stable magnetic environment. The movement of metal objects around the house or nearby – driving the car into the garage, for example – can give spurious readings, but true magnetic activity is distinctive.

Joining forces

Astronomers, like galaxies, frequently gather in local groups. If you live in a big enough town, you may find that there is a club or society that meets near you. Here you will discover the wide range of amateur types – the experienced observers, the eclipse chasers, the armchair astronomers, the telescope nuts, and the out-and-out loonies. But most will be ordinary folk like yourself, probably without much scientific background, and simply keen to find out more about astronomy. Some will be more advanced than others, but do not worry that your own knowledge is inadequate. A good society caters for many levels and tastes. Give yourself a few meetings to fit in, for your first visit might coincide with the most technical or the worst talk of the year.

Most societies will have members who can help you with buying equipment, or advise on astrophotography, for example. You may even find that you have some particular knowledge or experience that enables you to help others. I have been a member of many societies, and have helped run more than a few. They can be fun and rewarding, and I would encourage anyone who wants to get more out of the subject to join one. But some would rather just get on with astronomy. Many people are by nature not joiners, in which case, fine.

There are larger organizations as well. Most countries have national amateur societies, such as the British Astronomical Association and the Society for Popular Astronomy (until recently called the Junior Astronomical Society) in Britain, the Royal Astronomical Society of Canada, and the Royal Astronomical Society of New Zealand. In the US there is no single organization, though the Astronomical League coordinates the activities of local societies, a task performed in Britain by the Federation of Astronomical Societies. The US has specialist organizations such as the Association of Lunar and Planetary Observers (ALPO) and the American Association of Variable Star Observers (AAVSO), both of which have members around the world, including professionals. In Australia there is the National Association of Planetary Observers.

There are also societies which, though based in one place, have an international membership. The Webb Society, devoted to deep-sky observing, is one of these. Another is The Astronomer, actually the name of a monthly magazine, though its subscribers make up an informal network of observers who often provide an invaluable service in checking new discoveries before they are officially announced. The magazine simply lists observations made the previous month, with occasional articles on observing techniques, while a circular service available by post, e-mail, and fax gives timely warning of a wide range of discoveries. The International Amateur–Professional Photoelectric Photometry group and the International Meteor Organization deal with their own respective areas of observing. Societies such as these do have meetings, but their value is that they provide a means of communication rather than acting as a social club.

Many local societies have their own observatories, some of which are equipped with meeting rooms, libraries, and even overnight accommodation. Rather than spend a large sum of money equipping yourself with a large telescope and all its ancillary equipment, you can join the society and use its telescope instead. But I would always advise having your own telescope, even if it is just a basic 150-mm (6-inch) reflector. The best telescope is the one that gets used the most.

The electronic society

A new form of astronomical society has grown up in recent years, without even meaning to. This is what you might call the electronic society. Its members chat regularly, though they may never meet face to face. They do so through modems attached to their computers.

Astronomy is a subject which lends itself to this form of communication. Even if you never compute anything, a computer is a handy thing to have. There is plenty of software which will show you the night sky in greater detail than you could ever hope to see through your telescope, in the form of programs which depict the sky, allow you to zoom in on parts of it, and display images and data. Other programs let you play with planets or galaxies, making them interact, or playing their motions far into the future or back into the past. With appropriate training you can design optical systems using ray-tracing programs. And once you have the computer, it is fairly cheap and simple to add a modem, which allows you to communicate with the outside world.

There are a few misconceptions about using modems. Some people believe that as soon as you connect with another number your computer is likely to become infected with a killer virus. Others think that you need a separate phone line. Neither of these is true. A modem is a means of converting the signals used by your computer into a form suitable for transmission over the telephone network. The easiest type of information to send is text, in the form of messages. It is highly unlikely that anyone could send you a message that could damage your computer in any way. Neither can people gain access to information stored on your computer, unless you choose to make it available to them. As for the phone line, you simply plug the modem into a standard telephone socket.

You could call friends who have modems and send messages to them directly, but it is more usual to use an intermediary. Some of these are free to use, while others charge a subscription. The free ones are generally what are called bulletin board systems (BBSs), set up by enthusiasts. They provide a host computer with modem and the appropriate software to run the board. To use one, you call the board's number via your computer, then after logging on (which usually consists of you giving your name and a password) you can either help yourself to information provided by the host, or leave a message for another user. When they log on, they find your message in their electronic mailbox.

There are a number of BBSs devoted to astronomy, and on a good one you can find an enormous amount of information which you can download (copy to your own computer). There may be the latest NASA press releases, articles on choosing telescopes, reviews of books and software, or a library of image files from spacecraft and of programs of interest. There are usually also "conferences" – open messages from other users. Frequently, for example, one person asks a general question (rather than sending the message to a specific person, they send it to "All"). Others who have opinions on the subject then reply. The messages are usually quite direct and chatty. Following a thread of conversation on a BBS is rather like being able to listen in to all the conversations at a society meeting without having to take part yourself.

The best way to follow what is happening on a bulletin board is to log on every few days and download the messages that sound interesting from their headings, which you can scan quickly. A program known as an "offline mail reader" simplifies and automates the process, reducing your phone bill. Then you can read the messages when you want to, and maybe compose a reply. The next time you log on, you upload (send) your replies. I frequently log on to an astronomy bulletin board operated from London, StarBase One, and find that I can collect a week's messages in several conference areas in just a couple of minutes.

One American system, CompuServe, has become used worldwide. Though the host computer is in Columbus, Ohio, it has direct-dial numbers in many countries so that your call may be charged only at local rates. The system looks the same on screen whether you are in Columbus or Copenhagen. Outside office hours the system is comparatively cheap to use, and you can join in any of the many forums devoted to special interests. The Astroforum is a very lively conference, with sometimes hundreds of messages left daily on a wide range of topics. Many of the first-hand quotes from amateurs in this book were the result of asking questions in the CompuServe Astroforum. Some telescope manufacturers and retailers have discovered that it is a good way to find out what people really think about their products, and if you are considering buying a particular instrument you can simply post a message there asking if anyone has experience of using it.

If you feel you are not ready to pitch in yourself, you can remain a "lurker" for as long as you like. But, as with a regular society, you will find that there are many people just like you – and no one can see your red face if you do get things wrong! Although bulletin boards are not societies as such, they offer many of the same advantages and have a few more besides – you can often learn of astronomical discoveries long before they get into print.

Astronomical computing

At one time, devoted astronomers whiled away the cloudy nights by calculating such things as planetary positions for the coming years. These tasks are now largely carried out by computer, and there is little point in amateurs spending their time doing this. But there is still plenty to keep the enthusiast occupied. If you want to help out, you could offer to help analyse data obtained by other amateurs, such as variable-star observers, particularly if you have a computer. For the more adventurous, there is the possibility of developing your own software for image processing or display. Astronomy is becoming increasingly computer-oriented, and by writing software you can make your mark just as effectively as if you were an observer. This is not the sort of task at which you can become proficient overnight, but many societies have computer groups that will help get you started. *Sky & Telescope* carries a monthly column which explores the range of possibilities in programming a computer.

An increasing amount of astronomical information is now becoming available on CD-ROM, and the cost of obtaining a CD drive for your home computer is also dropping. The raw images obtained by NASA spacecraft are being made available in this format. Many of them have never been analysed, and it is now quite possible for amateurs to make a contribution in this field. Such work is best done under the guidance of a planetary geologist. Images from the Hubble Space Telescope are also being released on CD-ROM. As more and more data become readily available, it is becoming impossible for professionals to keep up with the analysis. There is real potential here for the amateur to make some contribution.

The North Polar Sequence

Visual magnitudes 12h Scale 4'.6 = 1 mm

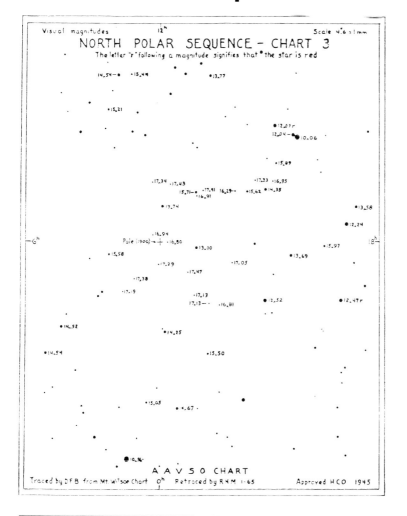

NORTH POLAR SEQUENCE – CHART 3

The letter "r" following a magnitude signifies that the star is red

A A V S O CHART

Traced by D F B from Mt. Wilson Chart 0h Retraced by R M M 1-65 Approved HCO 1945

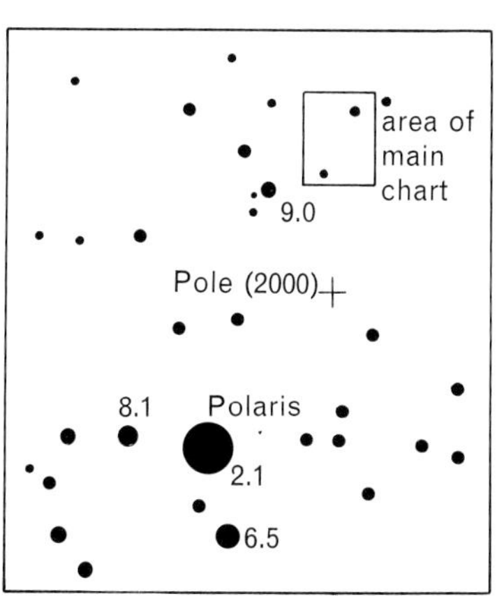

The North Polar Sequence – an area of stars close to the north celestial pole with accurately measured magnitudes. Use it to find the faintest star visible with your instrument, and to test the accuracy of your own photometric measures. The position of the pole has shifted slightly since the sequence was devised, but it remains at a virtually constant altitude throughout the year for northern hemisphere observers. No similar sequence is available for the southern hemisphere.

Deep-sky objects visible from the city

Unlike most other lists of deep-sky objects, these have been compiled especially for the city observer, with the help of David Frydman, Stanley Gleeson, Graham Long, Ian Ridpath, Steve Smith, and Richard Westwood. All the objects listed have been seen from city skies. Some may require ideal conditions and a fairly large aperture, and others may be beyond your reach. On the other hand, you may be able to see objects not listed here. Some general visibility guidelines are:

Any open cluster brighter than overall magnitude 7 is probably well seen if it is above 20° elevation.

Any globular cluster brighter than magnitude 8 is well seen if above 45° elevation.

Planetary nebulae can be seen if they are small and above 45° elevation, particularly if you use a filter. Some need a high magnification, which makes them hard to locate.

Galaxies are usually difficult from cities unless they are higher than 65°, especially if they are seen face-on rather than edge-on. Filters do not usually help.

Few diffuse nebulae are well seen from towns, though a filter can improve their visibility.

Double stars are well seen above 20°.

The larger your telescope, the fainter the objects you can see – even from bright locations.

The top ten city objects

No definitive choice is possible, because even bright objects can be invisible when low in the city sky. Observers in Britain and Canada get a poor view of M6 and M7, for example, while an observer farther south would rank M13 a very poor second to Omega Centauri. So I have divided the sky into three bands: high northern latitudes, including much of northern Europe and Canada; mid-northern latitudes, including the southern US and farther south; and the southern hemisphere. I have included some objects just because people want to see them, rather than because they are particularly spectacular. All these objects should be observable from city skies without the use of a filter. I have put them roughly in order of interest, although the choice is a little subjective and others will have their own ideas.

1. Top ten showpieces visible from high northern latitudes

Designation	RA	dec	Object	Constellation	Comments
M45	$03^h\ 47^m$	+24° 07'	Loose cluster	Taurus	Pleiades. Needs a low power on a telescope. Excellent in binoculars
M42	05 35	−05 30	Diffuse nebula	Orion	Orion Nebula. Visible with the naked eye even from the city; gets better the more aperture you can use
NGC 869 and 884	02 20	+57 08	Open clusters	Perseus	Double Cluster/h and Chi. Naked-eye object when high up; low power or binoculars. Look for the red supergiant stars in NGC 884
M44	08 40	+20 00	Open cluster	Cancer	Praesepe or Beehive. Naked-eye and binocular object
M27	20 00	+22 43	Planetary nebula	Vulpecula	Dumbbell Nebula. Can just be seen with binoculars. Its dumbbell shape is clear with a power of about 20
M57	18 54	+33 02	Planetary nebula	Lyra	Ring Nebula. Needs a power of 50 or more to show it, but a favourite object for small telescopes
Beta Cygni	19 31	+27 58	Double star	Cygnus	Albireo. Contrasting colours make this one of the sky's best doubles – needs a power of 20 or more
M31	00 43	+41 16	Galaxy	Andromeda	Andromeda Galaxy. The brightest galaxy; visible to the naked eye and worth a look with any optical aid you have. Not spectacular, but a favourite
M13	16 42	+36 28	Globular cluster	Hercules	Barely visible in binoculars – you may mistake it for a star. Use a moderate power on a telescope to resolve stars at the centre

2. Top ten showpieces visible from mid-northern latitudes

Designation	RA	dec	Object	Constellation	Comments
M6	17^h 40^m	−32° 13'	Open cluster	Scorpius	Hard to see from Britain and Canada
M7	17 54	−34 49	Open cluster	Scorpius	Hard to see from Britain and Canada
M8	18 04	−24 23	Diffuse nebula	Sagittarius	Lagoon Nebula. Visible in binoculars
M11	18 51	−06 16	Open cluster	Scutum	Wild Duck. Visible in binoculars, but needs more aperture for best views
NGC 253	00 48	−25 17	Galaxy	Sculptor	Edge-on spiral. Hard to see from Britain and Canada
Omega Centauri	13 27	−47 29	Globular cluster	Centaurus	Good in binoculars; gets better with increasing aperture. Best seen from the southern hemisphere
M17	18 21	−16 11	Diffuse nebula	Sagittarius	Omega Nebula
M41	06 47	−20 44	Open cluster	Canis Major	Good in binoculars
NGC 6231	16 54	−41 48	Open cluster	Scorpius	Best seen from the southern hemisphere
NGC 2477	07 52	−38 33	Open cluster	Puppis	Best seen from the southern hemisphere

3. Top ten southern showpieces

Designation	RA	dec	Object	Constellation	Comments
Large Magellanic Cloud (LMC)	05^h 30^m	−69°	Galaxy	Dorado	Naked-eye and binocular object
Small Magellanic Cloud (SMC)	01 00	−72	Galaxy	Tucana	Naked-eye and binocular object
47 Tucanae	00 24	−72 05	Globular cluster	Tucana	Very close to SMC
NGC 3372	10 45	−59 50	Diffuse nebula	Carina	Eta Carinae Nebula. Naked-eye object, includes many stars
NGC 2070	05 39	−69 06	Diffuse nebula	Dorado	Tarantula Nebula. In LMC. Bright and complex, enhanced by filter
NGC 3532	11 06	−58 40	Open cluster	Carina	
IC 2602	10 43	−64 24	Open cluster	Carina	Southern Pleiades. Needs low power and wide field
NGC 4755	12 54	−60 20	Open cluster	Crux	Jewel Box. Disappointing in binoculars – small number of stars. Needs some aperture and power to show a brilliant range of star colours
NGC 2516	07 58	−60 52	Open cluster	Carina	
NGC 3132	10 38	−40 26	Planetary nebula	Vela	Brighter than the Ring Nebula; needs similar power. Prominent central star

4. Top ten photographic objects

These are all comparatively easy to photograph using a filter and telephoto lens, yet virtually impossible to see from the city without a filter. In addition, many other diffuse nebulae mentioned in the lists on these pages can be photographed.

Designation	RA	dec	Object	Constellation	Comments
NGC 7000	21^h 00^m	+44° 20'	Diffuse nebula	Cygnus	North America Nebula. Popular object, but needs good red sensitivity and contrast to show it well
NGC 2237	06 30	+05 03	Diffuse nebula	Monoceros	Rosette Nebula. An easy object
NGC 1499	04 01	+36 37	Diffuse nebula	Perseus	California Nebula. Large and faint
B33, IC 434	05 41	−02 24	Dark nebula	Orion	Horsehead Nebula. Needs a long focal length to reveal the actual Horsehead, B33, though the bright nebula IC 434 can be photographed with a standard lens
NGC 2024	05 42	−01 51	Diffuse nebula	Orion	Flame Nebula. Just east of Zeta Orionis. Visible in telescopes using filter on a good night
NGC 6960/92	20 56	+31 43	Supernova remnant	Cygnus	Veil Nebula. The two nebulae form arcs of a circle. Position given here is for 6992, the brightest part. Visible on good nights with filter
NGC 2264	06 41	+09 53	Cluster, dark nebula	Monoceros	Cone Nebula. As with the Horsehead, needs long focal length to show the dark nebula
–	05 50	00 00	Supernova remnant	Orion	Barnard's Loop. Faint, but shows up because of its size (about 10°)
NGC 281	00 53	+56 37	Diffuse nebula	Cassiopeia	Many other nebulous objects to be photographed in Cassiopeia
–	20 22	+40 15	Diffuse nebulae	Cygnus	Gamma Cygni. Whole area wreathed with nebulosity

5. Top ten double stars

Star	RA	dec	Magnitudes	Comments
Beta Cygni	$19^h\ 31^m$	+27° 58'	3.1, 5.1	Albireo
Gamma Andromedae	02 04	+42 20	2.3, 5.1	Almach
Epsilon Lyrae	18 44	+39 40	4.7, 4.6	The Double Double
Theta Orionis	05 35	−05 23	5.1, 6.7	Trapezium; the two other stars are variable
Epsilon Boötis	14 45	+27 04	2.5, 4.9	Pulcherrima
Alpha Crucis	12 27	−63 06	1.3, 1.7	Acrux
Alpha Centauri	14 40	−60 50	0.0, 1.3	Rigil Kentaurus
Sigma Orionis	05 39	−02 36	4.0, 6.0	
Eta Cassiopeiae	00 49	+57 49	3.5, 7.5	
Delta Cephei	22 29	+58 25	3.6, 6.3	The original Cepheid star. The primary varies from magnitude 3.6 to 4.3

6. Now try these

Having whetted your appetite on the sky's showpieces, you will want to look further. All the objects in this list are visible from city locations; there are many others besides, particularly in the southern sky. Many of these objects can be picked out with binoculars, but you need a telescope to do them proper justice – and the larger it is, the better the view.

Designation	RA	dec	Object	Constellation	Comments
NGC 55	$00^h\ 15^m$	−39° 11'	Galaxy	Sculptor	
M110	00 40	+41 41	Galaxy	Andromeda	See comments for M32. Also known as NGC 205
M32	00 43	+40 52	Galaxy	Andromeda	Famous mostly as a companion to M31, otherwise unimpressive
NGC 404	01 09	+35 43	Galaxy	Andromeda	Close to Beta Andromedae, which must be kept out of the field of view. A test of the contrast of your optics
NGC 457	01 19	+58 20	Open cluster	Cassiopeia	Owl Cluster. Fine contrast of colour, with a fairly compact centre
M103	01 33	+60 42	Open cluster	Cassiopeia	
M33	01 34	+30 39	Galaxy	Triangulum	Pinwheel Galaxy. Very difficult in light-polluted skies
M76	01 42	+51 34	Planetary nebula	Perseus	Little Dumbbell. Needs a high power
NGC 752	01 58	+37 41	Loose cluster	Andromeda	Large but unimpressive with binoculars; low power on larger telescope needed
M34	02 42	+42 47	Open cluster	Perseus	Includes several pairs and triple stars
M77	02 43	−00 01	Galaxy	Cetus	
Melotte 20	03 20	+50	Loose cluster	Perseus	Alpha Persei cluster
NGC 1502	04 08	+62 20	Open cluster	Camelopardalis	Has a group of white stars in the shape of a box
NGC 1514	04 09	+30 47	Planetary nebula	Taurus	Filter needed
NGC 1528	04 15	+51 14	Open cluster	Perseus	
–	04 30	+17	Loose cluster	Taurus	Hyades. Too large (about 5°) to be seen in any instrument but binoculars
NGC 1647	04 46	+19 04	Open cluster	Taurus	Needs low power on a large aperture for the best views. Near the Hyades
NGC 1758	05 04	+23 46	Open cluster	Taurus	Very large cluster
M38	05 29	+35 50	Open cluster	Auriga	Has a greater mix of bright and faint stars than M34 or M36. Look for the fainter NGC 1907 near its edge
M1	05 35	+22 01	Supernova remnant	Taurus	Crab Nebula. Needs moderate power Not spectacular visually without a filter
NGC 1981	05 35	−04 26	Open cluster	Orion	
M36	05 36	+34 08	Open cluster	Auriga	Together with M37 and M38, makes a fine sight in binoculars
NGC 1977	05 36	−04 52	Diffuse nebula	Orion	
M43	05 36	−05 16	Diffuse nebula	Orion	Northeast part of Orion Nebula
NGC 2017	05 39	−17 51	Multiple star	Lepus	
M78	05 47	+00 03	Diffuse nebula	Orion	Comet-like
M37	05 52	+32 33	Open cluster	Auriga	Needs darker sky than M36 and M38
M35	06 09	+24 20	Open cluster	Gemini	Compact cluster NGC 2158 is nearby, visible in larger telescopes
NGC 2244	06 32	+04 52	Open cluster	Monoceros	Surrounded by the Rosette Nebula, NGC 2237 (see photographic list), which is visible using a filter
NGC 2251	06 35	+08 22	Open cluster	Monoceros	Linear star patterns
NGC 2264	06 41	+09 53	Open cluster	Monoceros	

M50	$07^h\ 03^m$	−08° 20'	Open cluster	Monoceros	
NGC 2362	07 19	−24 57	Open cluster	Canis Major	Needs moderate power
NGC 2392	07 29	+20 55	Planetary nebula	Gemini	Eskimo Nebula, so called because of apparent features within the nebula
M47	07 37	−14 30	Open cluster	Puppis	
M46	07 42	−14 49	Open cluster	Puppis	
NGC 2451	07 45	−37 58	Open cluster	Puppis	
M67	08 50	+11 49	Open cluster	Cancer	
NGC 2903	09 32	+21 30	Galaxy	Leo	One that Messier missed
M81	09 56	+69 04	Galaxy	Ursa Major	A binocular object in good skies. In the same field as M82
M82	09 56	+69 41	Galaxy	Ursa Major	Better from city skies than nearby M81
NGC 3242	10 25	−18 38	Planetary nebula	Hydra	The Ghost of Jupiter, so called because of its size and elliptical shape
M65	11 19	+13 05	Galaxy	Leo	Noticeably elliptical
M66	11 20	+12 59	Galaxy	Leo	Less tilted than M65. Nearby stars make M65 and M66 easy to find
NGC 3918	11 50	−57 11	Planetary nebula	Centaurus	A small but bright planetary, blue in colour
Melotte 111	12 25	+26	Loose cluster	Coma Berenices	Coma Star Cluster. Like the Hyades, too large for a telescope
M104	12 40	−11 37	Galaxy	Virgo	
NGC 4945	13 05	−49 28	Galaxy	Centaurus	Very faint from city site; a large, long object
M53	13 13	+18 10	Globular cluster	Coma Berenices	
NGC 5128	13 26	−43 01	Galaxy	Centaurus	Centaurus A. A large, slightly oblong blob with just a hint of a dust lane
M3	13 42	+28 23	Globular cluster	Canes Venatici	
NGC 5307	13 51	−51 12	Planetary nebula	Centaurus	Difficult, very small and not much larger than a star. Filter needed
M101	14 03	+54 21	Galaxy	Ursa Major	
IC 4406	14 22	−44 09	Planetary nebula	Lupus	Small object, relatively bright and circular
M5	15 19	+02 05	Globular cluster	Serpens	
NGC 5986	15 46	−37 47	Globular cluster	Lupus	Small object, fairly faint. One star at its edge is relatively bright
M4	16 24	−26 32	Globular cluster	Scorpius	
NGC 6210	16 45	+23 49	Planetary nebula	Hercules	
M12	16 47	−01 57	Globular cluster	Ophiuchus	
M10	16 57	−04 06	Globular cluster	Ophiuchus	
M92	17 17	+43 08	Globular cluster	Hercules	
M14	17 38	−03 15	Globular cluster	Ophiuchus	
NGC 6397	17 41	−53 40	Globular cluster	Ara	Easy object, a fairly open cluster but with condensed nucleus
IC 4665	17 46	+05 43	Loose cluster	Ophiuchus	Best seen in binoculars – a loose scattering of stars
NGC 6543	17 59	+66 38	Planetary nebula	Draco	
M16	18 19	−13 47	Cluster and nebula	Serpens	Eagle Nebula
NGC 6633	18 28	+06 34	Open cluster	Ophiuchus	Needs a low power; bright, yellowish stars
M22	18 36	−23 54	Globular cluster	Sagittarius	
M56	19 17	+30 11	Globular cluster	Lyra	
Collinder 399	19 25	+20 11	Loose cluster	Vulpecula	Coathanger. Also known as Brocchi's Cluster
NGC 6826	19 45	+50 31	Planetary nebula	Cygnus	The Blinking Planetary, so called because it seems to blink on and off as the eye is attracted to the bright central star
M71	19 54	+18 47	Globular cluster	Sagittarius	
M29	20 24	+38 32	Open cluster	Cygnus	Needs low power
IC 5067	20 48	+44 22	Diffuse nebula	Cygnus	Pelican Nebula. Needs filter
NGC 7006	21 01	+16 11	Globular cluster	Delphinus	
NGC 7009	21 04	−11 22	Planetary nebula	Aquarius	Saturn Nebula
NGC 7027	21 07	+42 14	Planetary nebula	Cygnus	
M15	21 30	+12 10	Globular cluster	Pegasus	
M39	21 32	+48 26	Open cluster	Cygnus	
M2	21 34	−00 49	Globular cluster	Aquarius	
NGC 5150	21 59	−39 25	Planetary nebula	Grus	Very faint, reasonable size. Filter helps
NGC 7293	22 30	−20 48	Planetary nebula	Aquarius	Helix Nebula. Needs filter
NGC 7331	22 37	+34 25	Galaxy	Pegasus	
NGC 7510	23 12	+60 34	Open cluster	Cepheus	Strange bar shape
M52	23 24	+61 35	Open cluster	Cassiopeia	Similar to M37, but maybe easier
NGC 7662	23 26	+42 33	Planetary nebula	Andromeda	Needs high power
NGC 7789	23 57	+56 44	Open cluster	Cassiopeia	The higher in the sky and the larger the telescope, the better it gets

Bibliography

Atlases and catalogues

Burnham, R. Jr, *Burnham's Celestial Handbook*, Dover, 1978.

Lovi, G., and Blow, G., *Monthly Star Charts*, Sky Publishing Corp., 1994.

Luginbuhl, C.B., and Skiff, B.A., *Observing Handbook and Catalogue of Deep-Sky Objects*, Cambridge University Press, 1989.

Ridpath, I. (editor), *Norton's 2000.0*, Longman/Wiley, 1989.

Sinnott, R. W. (editor), *NGC 2000.0*, Cambridge University Press/Sky Publishing Corp., 1988.

Thompson, G.D., and Bryan, J.T. Jr, *Supernova Search Charts and Handbook*, Cambridge University Press, 1989.

Tirion, W., *Sky Atlas 2000.0*, Sky Publishing Corp., 1981.

Manuals on general amateur astronomy and observing guides

Berry, R., *Build Your Own Telescope*, second edition, Scribner's, 1994.

Berry, R., *Choosing and Using a CCD Camera*, Willmann-Bell, 1992.

Covington, M.A., *Astrophotography for the Amateur*, revised edition, Cambridge University Press, 1991.

MacRobert, A., *Star-Hopping for Backyard Astronomers*, Sky Publishing Corp., 1993.

Muirden, J. (editor), *Sky Watcher's Handbook*, W.H. Freeman, 1993.

Newton, J., and Teece, P., *The Guide to Amateur Astronomy*, Cambridge University Press, 1988.

Publications

Astronomy, Kalmbach Publishing, 21027 Crossroads Circle, P.O. Box 1612, Waukesha, WI 53187, USA (monthly).

Astronomy Now, Hall Park Publications Ltd, Douglas House, 32–34 Simpson Road, Bletchley, Milton Keynes MK1 1BA, UK (monthly).

CCD Astronomy, Sky Publishing Corp., P.O. Box 9111, Belmont, MA 20178-9111, USA.

Earth at Night poster (page 8), Hansen Planetarium, 1845 South 300 West #A, Salt Lake City, UT 84115, USA; available in the UK from Armagh Planetarium, College Hill, Armagh, BT61 9DB, UK.

Observer's Handbook, Royal Astronomical Society of Canada, 136 Dupont Street, Toronto, Canada M5R 1V2 (annually).

Sky & Telescope, Sky Publishing Corp., P.O. Box 9111, Belmont, MA 02178-9111, USA (monthly).

Useful addresses

Organizations

American Association of Variable Star Observers, 25 Birch Street, Cambridge, MA 02138-1205, USA.

Association of Lunar and Planetary Observers, 8930 Raven Drive, Waco, TX 76712, USA.

Astronomical League, 14601 55th Street South, Afton, MN 55001, USA.

British Astronomical Association, Burlington House, Piccadilly, London W1V 9AG, UK.

Campaign for Dark Skies, c/o British Astronomical Association.

Federation of Astronomical Societies (UK), Whitehaven, Maytree Road, Lower Moor, Pershore, Worcestershire WR10 2NY, UK.

International Amateur–Professional Photoelectric Photometry, 629 North 30th Street, Phoenix, AZ 85008, USA.

International Dark-Sky Association, 3545 N. Stewart, Tucson, AZ 85716, USA.

International Meteor Organization, Pijnboomstraat 25, B-2800 Mechelen, Belgium.

International Occultation Timing Association, 2760 SW Jewell Avenue, Topeka, KS 66611-1614, USA.

National Association of Planetary Observers, P.O. Box 2, Riverwood, NSW 2096, Australia.

Society for Popular Astronomy (formerly the Junior Astronomical Society), 36 Fairway, Keyworth, Nottingham NG12 5DU, UK.

The Astronomer, 250 Linnet Drive, Chelmsford, Essex CM2 8AJ, UK.

Webb Society, 194 Foundry Lane, Freemantle, Southampton, Hampshire SO1 3LE, UK
  Secretary North America: 1440 S. Marmora, Tucson, AZ 85713-1015, USA.
  Southern Secretary: P.O. Box 40, Manly, Queensland 4179, Australia.

Manufacturers

Astro Cards, P.O. Box 35, Natrona Heights, PA 15065, USA.

Astro-Physics, Inc., 11250 Forest Hills Road, Rockford, IL 61115, USA.

Celestron International, 2835 Columbia Street, Torrance, CA 90503, USA.

Dark Star Telescopes, 6 Pinewood Drive, Ashley Heath, Market Drayton, Shropshire TF9 4PA, UK.

Electrim Corporation, 353 Nassau Street, Princeton, NJ 08540, USA.

Hale Research, Cinderford, Gloucestershire GL14 2QW, UK.

Lumicon, 2111 Research Drive #5S, Livermore, CA 94550, USA.

Meade Instruments, 16542 Millikan Avenue, Irvine, CA 92714, USA.

Orion Optics, Unit 12, Quakers Coppice, Crewe Gates Farm Industrial Estate, Crewe CW1 1FA, UK.

Orion Telescope Center, 2450 17th Avenue, PO Box 1158-S, Santa Cruz, CA 95061-1158, USA.

Santa Barbara Instrument Group, 1482 East Valley Road, Suite #J601, Santa Barbara, CA 93108, USA. (ST-4 and ST-6 CCD cameras)

SpectraSource Instruments, 31324 Via Colinas, Suite 114, Westlake Village, CA 91362, USA. (Lynxx CCD camera)

Starlight Xpress, September Cottage, Murrell Hill Lane, Binfield, Berkshire RG12 5DA, UK.

Tele Vue Optics, Inc., 100 Route 59, Suffern, NY 10901, USA.

Amateur observing sites

Association Newton 406, La Remise, 04700 Puimichel, France.

Centre for Observational Astronomy in the Algarve, Poio, Mexilhoeira Grande, 8500 Portimão, Portugal.

Star Hill Inn, Box 1A, Sapello, NM 87745, USA.

Electronic bulletin boards

CompuServe, PO Box 20212, 5000 Arlington Centre Blvd, Columbus, OH 43220, USA.

StarBase One, telephone 0171–703 3593 or 0171–701 6914; modem setting 8-N-1, speeds up to 14,000 bps.

Index

Page numbers in *italic* refer to an illustration on a different page from the text. References to objects in the lists on pages 168–71 are not included.